AMERICAN MEDICINE AND
STATISTICAL THINKING,
1800–1860

American Medicine and Statistical Thinking, 1800–1860

James H. Cassedy

HARVARD UNIVERSITY PRESS

CAMBRIDGE, MASSACHUSETTS

LONDON, ENGLAND

1984

Library of Congress Cataloging in Publication Data
Cassedy, James H.
  American medicine and statistical thinking, 1800–1860.

  Includes bibliographical references and index.
    1. United States—Statistics, Medical—History—19th century. 2. Medical statistics—History—19th century. 3. Medicine—United States—History—19th century. 4. Medicine—United States—Statistical services—History—19th century. I. Title.
  RA407.3.C37   1984   362.1'072073   83–12831
  ISBN 0–674–02560–1

For
Karen and Laura

Preface

This volume deals with two discrete but related topics in American history. It is, for one, an examination of American medicine during the crucial early years of its formation as a self-conscious profession and science. It is equally a study of the introduction and rapid spread of applied statistical methods and modes of thinking in a large area of American intellectual and scientific life during the same period.

The extensive use of numerical analysis in antebellum medicine provides a convenient illustration of the coming-of-age of applied statistics in this country. In fact, the collection, study, and application of numerical data became one of the central characteristics of American medicine during this period. As such, it assumed a permanent major role in medical administration and medical investigation, just as it did in the organization and conduct of other professional and intellectual fields.

This book carries forward some of the lines of statistical and medical history that were begun in my *Demography in Early America* (1969). Specifically, within a broad context of nineteenth-century medical inquiry and development, I am interested here in the refinement of observational techniques, in the proliferation of data-collection activities, in the extension of statistical analysis as an intellectual tool, and in the beginnings of statistical institutions and professionalism. The terminal date 1860, while partly an arbitrary one aimed at keeping a huge body of material within manageable limits, is also a logical one, reflecting the high point of nineteenth-century American enthusiasm for statistics.

Interpreting American medicine through its numerical applications admittedly does not provide a balanced historical view. By focusing primarily upon numerically inclined physicians, I have restricted my attention to the most intelligent and articulate members of the profession, and to only limited aspects of their activities as physicians. My search for the quantitative aspects of medicine has left little room for bringing in the qualitative, for considering other components of medical research. My interest in tracing the numerical growth of institutions has left little opportunity for examining either the substantive history of medical schools, societies, and hospitals, or many of the human aspects of health care.

Although this volume deals with a considerable spectrum of statistical matters, it does not attempt to provide a systematic technical evaluation of the various modes or examples of early medical statistics that it discusses. Rather, it is concerned with portraying the statistically minded physicians of the period, examining the scope of their activities and determining their place in American medicine and American history. More often than not, however, the data of antebellum nineteenth-century medicine were crude and faulty, often scant and incomplete in quantity and suspect in quality. They were often carelessly collected, frequently representative of selected cases only, and rarely differentiated as to variables. They were presented naively, used by readers uncritically, and often mistaken for knowledge. In the main, they also constituted the simplest form of empirical statistics, a far cry from the sophisticated mathematical statistics that finally came into its own in the twentieth century.

I have attempted to discuss early nineteenth-century medical statisticians and their work at their own level and on their own terms. Faulty as their efforts often were, I have concluded, with most of these individuals, that their cumulative groping uses of numerical methods and statistical analysis were important if only as improvements over what came before. Members of the antebellum medical profession relied upon statistics as the essential basis for challenging outmoded theories, for justifying new medical institutions in their communities, and generally for bringing about health reforms. They regarded statistics as the symbol of medical progress and as the primary hope of attaining medical certainty in their time.

In American medicine as well as in society generally, nineteenth-century statistical activities and aspirations reflected a rapidly expanding concern for people collectively, a concern which made important inroads upon the dominant individualism of the times. In fact, as this concern became institutionalized, some of its spokesmen came to conceive of statistics as a separate science or branch of knowledge, one whose practitioners would devise the essential guidelines and controls for society. Although this remained only an ideal, it did result in a twofold use of the word *statistics*. In the plural it denoted accumulations of similar data; in the singular, a method, discipline, or form of thought. Along with the nineteenth-century medical writers, then, I use the word both ways, a practice which sometimes makes the word *statistical* virtually synonymous with *numerical* or *quantitative*.

The historian George Daniels pointed out in *American Science in the Age of Jackson* (Columbia University Press, 1968) that by the beginning of the nineteenth century, "as the early data-collecting stage grew to a close in a number of sciences, the [essentially empirical] Baconian method had begun to show signs of having exhausted its potentialities" (p. 118). True as this was of other sciences, it was far from applicable to medicine. In fact, under the name of "statistics," both the gathering of medical data and Baconian analysis of the data demonstrated consistent rapid growth from 1800 to 1860, a growth which went hand in hand with the dramatic general growth and expansion of the United States during the same period.

I am profoundly grateful to my colleagues at the National Library of Medicine for their long, continued interest and support, and to my supervisors at the library for providing me with the opportunities and facilities necessary for writing this book. Celia Volen and Margaret Donovan generously took large amounts of time from their other duties to type the manuscript in its various stages.

The formulation and pursuit of this study has benefited greatly over the years from the continuing interchange of ideas with Professors I. Bernard Cohen, Robert C. Davis, John M. Eyler, and Barbara G. Rosenkrantz. Drs. David V. Glass, Lester S. King, and Marion M. Torchia were kind enough to read and criticize early versions of several chapters. Drs. John B. Blake, Gert H. Brieger, and Robert C. Davis read the entire manuscript, made many helpful suggestions, and offered their encouragement. I am thankful to all of them.

J.H.C.

Contents

AMERICAN MEDICINE AND
STATISTICAL THINKING,
1800–1860

America's Early Health Inventory

Among the intellectual and administrative enterprises which engaged America's citizens after the Revolution, none was more sweeping or more urgent than that of taking stock of the new United States, making a factual determination or inventory of the nation's assets and liabilities. This was a partly deliberate and partly spontaneous effort to determine the cumulative strength of a collection of individual states that still clung to their essential autonomy. Americans needed information about land, trade, and physical resources so they could determine the nation's potential for wealth and happiness and get on with their everyday business. They wanted information about their recent history so they could celebrate among themselves and to the world the unique virtues of their republican land and government. Medically, they needed to know the extent of the diseases in their midst, as well as the personnel, institutions, and resources that could help them cope with them.

America's Founding Fathers recognized that, after Independence, their needs for information went far beyond those of colonial Americans under British rule. With no special coordination, the nation thus formulated and launched a wide range of data-gathering projects, some of them governmental, others private. The ultimate harvests of concrete usable facts that quickly began to appear were originally heavily impressionistic and expressed in largely descriptive terms. By the beginning of the nineteenth century, however, they were becoming increasingly quantitative and numerical. Certain indigenous precedents for both kinds of inquiries and reports existed from colonial experience, though the most authoritative models came from Europe.[1]

European Models for Fact-Finding

At the turn of the eighteenth century, American leaders and intellectuals were being influenced by the ideas of an impressive and diverse body of European savants who had brought a quantitative approach to social and scientific thought, or who otherwise advocated systematically collecting, organizing, and arranging facts for the conduct of human affairs. Two of the most important of these thinkers had founded a quantitative tradition in Britain that had become increasingly more vigorous during the preceding century and a half. In science, the name of Sir Francis Bacon had long been synonymous with the exhaustive amassing of facts and with rational decision-making based on inductions from them. In other areas of knowledge, moreover, Bacon's overall vision of an ordered social and political universe resting firmly on careful observation, data gathering, counting, and measuring seemed to be more popular than ever before.

By contrast, the "political arithmetic" of the pioneer statisticians John Graunt and William Petty had scarcely survived the seventeenth century as an actively used term. Nevertheless, it had founded a firm, permanent administrative heritage in the frequently numerical activities of Britain's imperial bureaucracy. Elsewhere, the quantitative outlook instilled by political arithmetic lived on, not only in the quiet world of scholars concerned with abstract mathematical probabilities, but far more conspicuously in the rapidly expanding commercial world of annuities and life insurance.

Political arithmetic's comprehensive, fundamentally numerical expression of large quantities of facts pertaining to society and the state was central to what the nineteenth and twentieth centuries came to call *statistics*. However, the latter term itself was not of British but of Continental origin. Throughout most of the eighteenth century, it was associated primarily with Gottfried Achenwall and certain other German political writers who, under the name of *Staatskunde* or *Statistik*, compiled descriptive factual accounts of the characteristics and workings of governments. Only in the last decade of the century was the term *statistics* given a numerical connotation and applied to other areas of interest, and only then did it begin to enter the vocabulary of the English-speaking world. Largely responsible for that shift was Sir John Sinclair, with his massive quantitative survey of Scottish life and institutions, *The Statistical Account of Scotland*. Sinclair attached the term *statistics* firmly to the numerical mode of fact-finding and factual presentation, and in so doing created a persuasive model for a host of similar productions on both sides of the Atlantic.[2]

In addition to such specific influences as the quantitative approach of Sinclair and the descriptive technique of Achenwall, turn-of-the-century intellectuals and leaders of the Western world were exposed to a general emphasis on orderliness in processing knowledge and affairs from a wide spectrum of the spokesmen of the Enlightenment—encyclopedists, classifiers, system makers, and codifiers—as well as from practical men of the world. The school of Scottish "common-sense" philosophy reinforced this emphasis with its insistence on the centrality of empirical experience. Concurrently, Jeremy Bentham expressed a highly quantitative goal for social science in his concept of the greatest happiness of the greatest number. And in France, the mindless violence of the Revolution was finally replaced by the Napoleonic emphasis on order, efficiency, and data collecting throughout the empire's social and administrative fabric.

Americans were reasonably well prepared to take advantage of this multifaceted injunction to describe, enumerate, and count, and they had the motivation to do so. To be sure, few Americans were sufficiently equipped to pursue the theoretical aspects of mathematical statistics. However, many did have sufficient education and enough simple arithmetic to enable them to adopt the other models of information gathering or processing and to apply them effectively to the various aspects of their national inventory. In pursuit of such ends they quickly made themselves masters of the survey and the questionnaire. They developed the talents needed to compile native geographies, histories, gazetteers, and directories. They proved energetic in founding colleges, academies, and libraries. They strongly supported the search for and diffusion of new knowledge about farming, science, and the practical arts, and they faithfully recorded the details of their explorations and domestic travels. Not least of all, they sharpened their information-gathering propensities with a proliferation of new newspapers and periodicals.

By and large, late eighteenth- and early nineteenth-century Americans became fully as enthusiastic, if not as adept, as Europeans in collecting, compiling, and consuming useful factual information, including numerical data. In fact, the freedom of private initiative that Americans enjoyed, together with the considerable autonomy of their state and local governments, probably enabled many more individuals to become involved in fact-finding and compilation activities than was the case in Europe. Certainly a large and diverse number of participants, laymen as well as physicians, contributed to the health-related portions of the early American inventory.

American Diseases about 1800

Turn-of-the-century American writers devoted much space to accounts of the country's healthfulness. They were not hesitant about volunteering their own impressions of this subject whenever necessary, but most of them tried their best to clothe those impressions with the authority of concrete fact.

Subjective opinion was especially prominent when writers attempted to convince their readers of the superior salubrity of a particular region. Given the primitive level of medical information, early nineteenth-century assertions such as the claim that New England was "the healthiest country in the United States" were clearly based more on rhetoric than on science and were easy for champions of other areas to oppose.[3] However, arguments for the exceptional salubrity of certain rural areas seemed unassailable and were rarely if ever challenged. The Reverend Samuel Williams, for instance, urged in 1794 that the palm for healthfulness belonged to Vermonters. No other people, he wrote, "have so few diseases, multiply so fast, or suffer so little from sickness. The disorders which wear away the inhabitants of wealthy cities, are almost unknown in the woods. Very few die, but under the unavoidable decays of nature, and the deaths are to the births, in no higher proportion than 1 to 4.8."[4]

Despite that rosy picture, rural Americans had ailments enough, health problems that were all too apparent to those who inquired about them. One who investigated such things most energetically was the peripatetic President of Yale College, Timothy Dwight. In his travels around rural New England and New York State, Dwight proved himself a worthy successor of the Reverend Ezra Stiles, not only as Yale's spokesman for religious orthodoxy, but as a compulsive counter of objects and collector of information. Supplementing his own observations with data from census reports, geographies, histories, and the like, he filled his diaries with information, partly descriptive and partly quantitative, on a multitude of topics, including diseases and health conditions.

By 1800, Dwight noted, such maladies as gout and "the stone" had become virtually unknown in New England, but the measles and influenza both appeared every few years. The principal diseases of the region were dysentery, typhus and scarlet fevers, pleurisy, "peripneumony," croup, cholera infantum, bilious and remittent fevers, chronic rheumatism, and pulmonary consumption. Most widely fatal was consumption. This was due, Dwight found, to the severity of the climate and the sudden changes of the weather, but also to a number of social factors, including intemperance; sedentary living; "*leaning forward* on the part of students, clerks, and several classes of mechanics"; and the wearing of too few and too thin garments by females. Dwight thought that the neoclassical revival had its limits in the United States, at least insofar as healthful dress was concerned: "A young lady dressed à la Grècque in a New England winter violates alike good sense, correct taste, sound morals, and the duty of self-preservation."[5]

Upper New York State, into which New Englanders had begun to flow by the thousands in the 1790s, had its own health problems and resources. About this time, individuals with chronic rheumatism and similar conditions were discovering the virtues of the area's mineral springs, particularly those at Ballston and Saratoga. Dwight first visited Ballston in 1792, when that tiny community was still a quiet and hardly known opening in the forest. Within twenty years, however, several thousand people annually were going there searching for relief from their complaints. "Within the last fifteen years," Dwight lamented, "the spot has become a favorite and fashionable watering place, not only for medical purposes, but still more for those of pleasure and dissipation." Before long, he feared, "all the scenes of dissipation which have been customarily exhibited at watering places in Europe will be annually repeated here."[6]

Dwight found that some common diseases, even consumption, appeared to be virtually nonexistent in upstate New York. In their place, between Utica and Buffalo anyway, he reported a broad distribution of goiter, caused, it appeared, by mineral deposits in drinking water. The worst problems in the areas newly opened, however, were the various combinations of fevers. The "fever and ague" was recognized as almost universal among immigrants soon after their arrival in the region and for several succeeding summers. Dwight noted that the health of new settlers there as elsewhere was affected by excessive exposure to the elements, along with such factors as "working a great part of their time in moist ground, the badness of their houses, the poorness of their fare, together with the difficulty of obtaining proper medicines, good nurses, and skillful physicians."

Many other parts of rural America, both north and south, seemed to be equally unhealthy. Benjamin Latrobe, architect for the new capitol in Washington, had various opportunities to confirm this in his travels during the first decade of the century. As he wrote, "In my trips to the quarries to look over the stone destined for the public buildings, I have remarked upon the hundreds of half-starved, miserably lodged, idle, besotted, and fever-smitten families that inhabit the country on the Potomac, and indeed all the back country of the slave States below the mountains."[7]

Travelers like Dwight and Latrobe could gain a general idea of the health of these sparsely settled regions without much difficulty. Anything more than rough impressions had to wait until a substantial reservoir of local medical facts and data could accumulate in a community, until there was significant growth in population, institutions, and learning. Observers in established cities in 1800 could thus normally base their impressions of salubrity upon fact more than rural observers could.

Benjamin Rush, for example, drew upon several concrete sources in preparing a commentary on the state of health in Philadelphia about 1800, but he still relied heavily upon his own hunches and impressions. He thought that conditions had been changing for the better in the city over the previous several decades. Changes in dress and fashions—the decline in the popularity of wigs, the increase in the use of umbrellas by both men and women, and the decreased popularity of tight dresses and stays—all must have had their effects. He wondered, moreover, if the round hats recently introduced among men could have caused the recent reductions in apoplexy and palsy. Rush also thought that health in Philadelphia had been improved by the elimination of "relishes and suppers," the increased consumption of vegetables, a decline in meat eating, and a decline in the drinking of alcohol, at least among the genteel. Other favorable influences included improved construction of houses, provision for the heating of churches, a decline in the "practice of sitting on porches exposed to the dew in summer evenings," and the increase of suburban living. Whether or not such changes had contributed to better health, several diseases that had been bothersome in the past had declined and were rarely found after 1800; these included dropsy, the "dry gripes," and scrofula.[8]

Rush's sources also revealed some unfavorable trends. Philadelphia's destructive fevers of the previous generation had not disappeared but were taking new forms. Gout, scarlet fever, cholera infantum, and hydrocephalus continued to be common. Liver inflammations had increased. Consumption remained prevalent, especially among women. Like Dwight, Rush thought that this was due to the wearing of light dresses which exposed the upper arms. "The frequency of consumptions from this cause," he wrote, "has given rise to a saying that 'the nakedness of the women, is the clothing of the physician.'" Rush also deplored the physiological effects of such personal habits as increased cigar smoking, excessive eating of ice cream, and drinking. And he noted the great increases in mental disorders brought on by the violent political passions of the day, as well as by the "many new and fortuitous modes of suddenly acquiring and losing property."[9]

Elsewhere, other observers drew up similar inventories— partly impressionistic and partly based on observed fact—of the diseases and health conditions in their respective cities. As the nineteenth century began, two diseases were worrisome above all others: yellow fever and smallpox. These worries gave rise to an exceptional number and variety of medical information-gathering projects.

By 1800 concerned physicians, editors, and officials in coastal cities already had gone to great lengths to gather information about yellow fever. Throughout the 1790s, communities had to cope with successive epidemics of this disease, though no one really understood its cause. To protect themselves they adopted such measures as removing various kinds of filth, imposing quarantines, and building bonfires of sulfur. They considered the strengthening and expanding of ways to obtain timely information throughout the epidemics essential to combating them. In the larger cities, boards of health appointed special messengers to collect daily information from physicians, hospitals, and sextons on new cases and deaths; they also solicited meteorological readings, issued lists of deaths or burials, identified the victims by state or country of origin, and located cases and deaths by street.[10]

After yellow fever epidemics had run their course, mayors canvassed the physicians of their communities to try to find out more about what had happened and why. Physicians who had been involved published summaries of the measures taken, together with data on cases and deaths, often to defend their own roles during the epidemic or their ideas about it.[11] Valentine Seaman, a doctor in New York City, took the initiative of supplementing the statistics of the 1798 epidemic there with a spot map that gave precise locations of cases and deaths, block by block.[12] Everywhere, in keeping with the etiological controversies of the day, "contagionists" believed that yellow fever had been imported from Santo Domingo or some other locus of infection, and they gathered data to support this view. "Anticontagionists," who believed in the fever's domestic origins in decaying filth, amassed their data with equal diligence. Judged solely by the sheer volume of facts collected in the 1790s, the anticontagionists had undoubtedly gained the dominant position in the medical world of the early 1800s. However, their data proved to be so inconclusive that, when yellow fever came back to southern ports in subsequent decades, the argumentation and fact-finding of the two schools of thought started up all over again.[13]

By contrast with the efforts directed against yellow fever, the early nineteenth-century attempt to control smallpox eventually yielded statistics that were highly conclusive. The introduction of smallpox vaccination into the United States just as the nineteenth century began called for as much scrutiny through statistics as New England's experiment with inoculation some eighty years earlier had. In fact, while Americans enthusiastically welcomed this major development against a dread disease, they left much of the initial burden of assessing it to Edward Jenner and his British colleagues. The Harvard medical professor Benjamin Waterhouse reported in 1800 on the vaccinations he had performed successfully on his son and others in his household, but his observations were not quantitative. They did include a historical account of Jenner's discovery and a description of the technique of vaccination. Two years later, Waterhouse reported the Boston Board of Health's successful public test of vaccination on nineteen volunteers, but he still did not provide figures on the wider use of the practice in Boston or elsewhere.[14] Reports from Europe, however, chronicled the rapid spread of vaccination and included figures on the remarkable decline in smallpox deaths, in London from 2,409 in 1800 to 622 in 1804, and in Vienna from 835 to 2.

The earliest quantitative reports of the American experience with vaccination came from urban physicians like John Redmond Coxe of Philadelphia and Valentine Seaman of New York. By 1807 Seaman had successfully vaccinated nearly eleven hundred people at the New York Dispensary, and many more by 1816. Looking at the city's bills of mortality for the fifteen years before 1801, he noted that during that prevaccination period, over one-tenth of the deaths had been from smallpox. Vaccination had changed this proportion drastically: "For the last twelve years, as appears by the reports of the corporation, 27,151 interments have been made; according to the former proportion, as many, at least, as 2715 of these would have died of smallpox; but as only 676 did die of that disease, we may fairly conclude that the practice of vaccination, with all its imperfections about it, has prevented within the last twelve years, upwards of 2000 of our fellow citizens from dying of the smallpox."[15]

The incidence of smallpox in the United States, as well as this country's initial experience with vaccination, followed that of Europe fairly closely. However, inquiries into the nation's health suggested a considerable number of diseases which appeared to be peculiarly American or which flourished to an unusual extent in the United States. The French observer C. F. Volney, during his three years in the country between 1795 and 1798, found four kinds of prevailing diseases that he declared were "the direct offspring of the soil and climate of this country": catarrhs and consumption, toothache and loss of teeth, intermittent fevers, and yellow fever. A few years later David Warden, drawing from medical authorities all along the Atlantic seaboard, agreed that some of these same conditions "distinguished the climate from that of Europe." He noted in particular that tooth decay or scurvy of the gums "has frightened the European traveller ever since it was noticed by [Peter] Kalm," but he thought that this disease should be attributed less to the climate than to Americans' extensive consumption of salt meat and warm tea.[16]

The British physician Anthony Fothergill noted that many of the diseases found in Philadelphia were common to Great Britain and Europe as well as to other parts of the United States, "yet in consequence of climate, situation, and modes of living assume a different character." The croup, for instance, rare in England and restricted there to children, in America was "not only frequent and fatal among children but also attacks adults." Both dropsy of the brain and cholera infantum were more widespread than in England, while convulsions and lockjaw were "peculiarly frequent and often fatal." Yellow fever obviously was not encountered in England.[17]

Americans generally conceded that some diseases acted in

distinctive ways in the American environment. In fact, it was not uncommon for physicians to assert, with a certain sense of proprietorship, as did Charles Caldwell, that Europeans could not hope to understand American diseases without extensive experience with them here. For yellow fever, as one observer noted, physicians staked out a particularly emphatic claim: "We true Americans consider this subject as our national property, and are jealous of any attempts among Europeans, especially Englishmen, to assume any control over it."[18]

At the same time, few if any Americans could be found who were willing to admit that the extremes and broad fluctuations of American climate were particularly deleterious to health. Fewer yet endorsed the assertions of Count Buffon, William Robertson, and other European critics that the American environment was generally conducive to disease and adversely affected longevity, sexual vigor, and bodily size. Although Jefferson had published a wealth of data in his *Notes on Virginia* to refute such charges, some early nineteenth-century Americans felt obliged to make their own rebuttals. A condescending English comment in 1812 that Massachusetts was "the country of epidemics, as much as it is of swamps, woods, and savannahs," clearly could not be allowed to pass without notice. A Boston reviewer appropriately replied that England was "the country of scrophula and consumption, as much as it is of fogs and vapours," while he marshaled two pages of statistical data to prove that America had a healthier climate and fewer diseases than England. Similarly, another commentator thought that, if Dr. Lyman Spalding's bill of mortality for Portsmouth was correct, Americans "may boast of a climate far more favourable to human life than any known."[19]

Studying the Physical Environment

Verification of such opinions and facts became an important concern of laymen and scientists alike. The medical profession, in fact, increasingly perceived that conclusive answers to these sweeping questions could not be expected in the isolated inquiries of a few individuals, even from book-length commentaries on American diseases by such energetic observers as William Currie or David Ramsay. America's emerging medical community thus made one of its first priorities the expansion of this aspect of the national inventory, to encourage the broadest possible collection of data about those features of the physical environment which were assumed to affecthealth.

Turn-of-the-century studies of the environmental factors bearing upon health and disease were aspects of a vigorous and continuing scientific tradition that was partly Baconian, partly neo-Hippocratic. On both sides of the Atlantic quantitative observations about the weather, climate, and topography were being collected in geographies, travel accounts, agricultural almanacs, botanical studies, histories, and other factual works produced by various members of the learned community. However, the systematic pursuit of the health-related aspects of such inquiries was clearly the province of the medical profession.

Many physicians were convinced that these investigations would ultimately reveal the causes of almost all infectious diseases, or at least the fevers. Although the inadequacies of mortality statistics made drawing accurate correlations between particular environmental conditions and specific causes of death difficult, individuals were not deterred from pursuing data related to those conditions. Since no one community could be expected to have the same combination of physical circumstances as another, investigators assumed that, theoretically, each should have its own environmental inquiry. The inquiries undertaken were not universally performed in a scientific manner. Yet, the large premium they placed on direct observation, description, and enumeration gave a predominantly scientific character to this part of the health inventory.

Some of America's fledgling medical societies began to stimulate medical-environmental inquiries early on. The Massachusetts Medical Society was encouraging such studies by the 1780s and 1790s. Almost as early, the Medical Society of South Carolina formalized a long tradition of encouraging medical-meteorological and topographical research in Charleston by requiring an annual "review of the weather and diseases of the current year," to be presented by one of the members. Statistical reports made to the society during the early 1800s under this provision by David Ramsay, Frederick Dalcho, Joseph Johnston, and others, were among the most detailed health reports for American cities before the Civil War.[20]

At about the same time, the Medical Society of the State of New York made efforts to obtain annual surveys of the "topography, geology, and mineralogy of any county in the state, together with an account of the prevailing diseases." David Arnell's report for Orange County in 1809 was one of the most thorough responses. Arnell presented an account of the local diseases together with thermometric readings and summaries of topographical features, geological peculiarities, mineral springs, and medicinal plants—details about this rural county's resources which had been largely unknown.[21]

A far more potent stimulus to the medical-topographical study of the United States, and to medical information-gathering generally, was the emergence of a new form of medical institution, the professional medical journal. Comprehensive and effective fact-finding clearly demanded the collective energies of many individuals and a way to publish the findings promptly and regularly. America's first successful medical journal was the *Medical Repository*, founded in New York in 1797 by Samuel Latham Mitchill, Elihu H. Smith, and Edward Miller. Decidedly Baconian in their approach, the editors used their journal to encourage systematic medical fact-finding. Originally concentrating on New York City, they gave as much attention to the city's bills of mortality, hospital statistics, data on prostitution and alcohol consumption, epidemics, and environmental characteristics as to individual case reports and reviews of medical and scientific literature. They also appealed to physicians and learned men in every part of the country to survey the diseases and medical-topographic features of their communities. Subsequently they published many articles from this larger constituency, accounts that were both descriptive and quantitative. The editors' hope was that the studies collectively would constitute a comprehensive profile of health in the United States, one providing an authoritative view of the causes and distribution of the country's diseases, together with information on effective methods of prevention and cure, on the comparative salubrity of given places, on the effects of the clearing and cultivation of land, and on the effects of civilized life itself.[22]

The physicians, naturalists, and others who ferreted out such information for the *Medical Repository* and for other journals were participants in an exciting era of national discovery. Roaming around the backcountry of the eastern seaboard, they frequently unearthed much more than the medical resources and facts they were looking for. Some came across scientific information of considerable importance. Some were strongly affected by the beauties of local rivers, hills, and lakes, and expanded their reports to describe them.[23] Such descriptions became parts of an emerging awareness of the aesthetic features of the American landscape. In the next generation this appreciation of nature found rich expression not only in the leading schools of American art and literature but in one of the fundamental lines of antebellum medical thinking.

At the same time, other observers focused predominantly upon the developing urban scene. Some of the physician-editors, in fact, along with others of an investigative bent, compiled comprehensive informational handbooks aimed at publicizing given cities. Among them, Samuel Latham Mitchill in 1807 drew together factual materials from various sources to prepare his *Picture of New York.* Something more than a city directory, this work was characterized as a "travellers' guide" to the city and was designed at least partly to attract new residents, particularly foreign immigrants. Similarly, in 1811, Dr. James Mease issued his *Picture of Philadelphia,* while Daniel Drake expanded his relatively modest *Notices Concerning Cincinnati* of 1810 into a full-fledged *Picture of Cincinnati* by 1815.[24]

The urban "picture" publications provided convenient summaries of the different local inventories at specific points in time. While they were frankly intended to project favorable images of a city's vitality, health, commerce, and potential for growth, they nevertheless were usually essentially accurate. At least, the health-related information they customarily contained—data on population, topography, geology, climate, and disease—tended to be fairly reliable. Often heavily quantitative, they were among the most characteristic early nineteenth-century exercises in political arithmetic.

Enumerating the "Subjects of Medical Practice"

It was clearly in America's best interest to obtain progressively better data on these matters. Most writers continued to agree that finding and compiling much of this information was essentially the responsibility of local governments, private organizations, and individuals. However, in determining how many potential patients there were—how many "subjects of medical practice"—the federal government, through its decennial census, did play an important role.

The census of 1790 was a relatively simple enumeration of population which provided quantitative information for several kinds of users. For the legislator it was the basis for apportionment of seats in Congress; for the economist it constituted a measure of growing national strength; for the humanitarian it provided data on the numbers and distribution of slaves needing to be educated, freed, or colonized in Africa. It did not enumerate deaths, births, or diseases.

At the turn of the century, literate men looked forward to updated information from the second census. Many hoped that it could be enlarged to illuminate such matters as the longevity, occupations, and countries of origin of the populace. Congress decided, however, to limit the census of 1800 to the same kinds of population data that had been obtained in 1790, deferring to later censuses more extensive inquiries into demographic and health-related matters.[25]

Congress's failure to provide for an enumeration of occupations left the country with little overall information on the numbers and distribution of medical practitioners, surgeons, apothecaries, or midwives. Moreover, in the absence of a national medical establishment, there was no alternate body which might have obtained such information. To be sure, local historians and compilers of directories often recorded details about individuals of assorted backgrounds and training who were engaged in the various healing arts in their communities. However, for years no one came up with reliable information or even educated guesses of how many of these individuals there were nationwide.

Until 1815 the paucity of medical institutions would have

made any national enumeration of such institutions an unprofitable exercise. For the most part only the larger cities could boast any kind of health-related facilities. However, they were slowly growing in number and scope under the pressures of increasing urban populations and as a result of civic anxieties stirred up by the yellow-fever epidemics of the 1790s. By 1800, Boston, Philadelphia, and New York, with their colleges, libraries, and museums, all had their almshouses, medical schools, and medical and scientific societies. Each also had drug manufacturers along with publishers who were marketing more and more medical literature. New York boasted a medical journal, and general hospitals existed there and in Philadelphia and New Orleans. Dispensaries and hospitals for the insane could be found in some communities.

Local historians and medical society orators celebrating the arrival of the nineteenth century made due note of the increasing number of health-care institutions, but hardly anyone seemed bothered that there were still so few. Even where such institutions existed, people seemed to take greater satisfaction in the knowledge that life in the republic was healthier than it was in monarchical Europe, and also healthier than it had been here in colonial times. Nevertheless, as people paid more attention to counting the deaths in the nation, they tended to modify some of their optimism if not also their rhetoric.

Records of Mortality

Like the counting of doctors and medical institutions, the recording of deaths was conducted throughout the early nineteenth century at local levels, by states, churches, societies, and private individuals. However, the coordination, continuity, uniformity, or enforcement of this activity that was necessary to provide anything approaching complete, accurate information in any given state was completely absent. Neither the civil registers of vital statistics in New England and a few other colonies nor the parish registers prevalent elsewhere had ever functioned very well, and both systems were disrupted further by the Revolution. Attempts to reestablish such arrangements in the newly independent states had little priority. Some states dropped their old arrangements completely and made no new ones for years. Some kept their old provisions for compulsory registration in effect but made no efforts to enforce them. New Jersey and a few other states compromised by making registration entirely voluntary.

Civil registration was long carried out almost exclusively for legal purposes. Seldom, if ever, were either the civil or the parish registers perceived to have any relation to medicine or public health. Only where bills of mortality were prepared, from parish or other records, did the vital records acquire any such interest.

By the turn of the century, sextons of Christ Church parish in Philadelphia had been publishing annual bills of mortality in broadside form for over fifty years.[26] These bills included figures on christenings and on specific causes of death in the parish, together with data on the burials in other cemeteries in the city. Since the bills were prepared by individuals with no medical training, their drawbacks became increasingly apparent over the years. Finally, in 1807 the city's board of health was empowered to compile its own mortality reports through returns obtained from physicians, clergymen, and others, and analyzed by medical men. These reports were among the earliest in the United States to reflect the actual numbers of deaths rather than the numbers of known burials.[27]

Both official and unofficial bills of mortality also began to be published in other American cities about 1800. In New England some of these were direct outgrowths of a tradition of vital records kept by clergymen as parts of their private journals. The more substantial lists, covering from ten to fifty years, became much sought-after by turn-of-the-century editors. Appearing in a wide variety of publications, these pastoral registers at the very least helped satisfy a morbid curiosity among the public. However, those which covered long periods of time and which detailed specific causes of death were increasingly used as bases for subsequent medical and scientific calculations. In 1791 sixty-two of these registers collected by the American Academy of Arts and Sciences from sources in Massachusetts and New Hampshire became the basis of the Reverend Edward Wigglesworth's pioneering table of life expectancies. While no one at the time analyzed the substantial medical content of these registers any further, the Academy's initiative did serve to stimulate additional interest around New England in the collection of vital data.[28]

Meanwhile Salem, Massachusetts, and other New England towns had occasionally published bills of mortality from figures of burials furnished by those in charge of graveyards. Boston newspapers had furnished such data to the public throughout the eighteenth century. However, that city did not regularize its official arrangements for issuing bills of mortality until 1810. At that time, in an act to regulate burials, the board of health was authorized to require sextons to provide weekly reports on all persons buried.[29] These simple numerical reports were issued as broadsides, which were subsequently often reprinted in the local press. They did not begin to approach the more detailed statistical health reports of Chester, Manchester, or Carlisle in England, or even the data of Philadelphia, but at least they provided Boston city officials with a simple basis for making public-health decisions.[30]

In New York City the key step in the collecting of mortality data was the creation of the office of City Inspector in 1804. This office, among its other functions, provided a focus for the internal sanitary activities of the city and for the gathering and publishing of vital statistics. These functions were recognized as complementary and were pursued with great vigor during the rest of the decade under the direction of reformer-philanthropist John Pintard.

Like officials in other communities, Pintard had to be satisfied with statistics that were far from adequate. Although New York physicians and midwives were now required to report births monthly, and the clergy to report marriages, few actually complied, so such data were very incomplete for many years. Obtaining a fairly complete weekly report of official mortality figures, however, did prove feasible. As in Boston, this depended on compulsory reports from the sextons or caretakers of the city's cemeteries.[31] Earlier, Noah Webster had found that few sextons had felt any obligation to keep such records, but Pintard, with the new law behind him, could gather enough data to get official current bills of mortality started and published regularly from 1804 on.[32]

Pintard's first summary of these data expressed shock at the high proportion of infant mortality in New York, along with the many deaths from consumption, drowning, and suicide. It went on to advise the community of the importance of continuing its mortality reports regularly: "The public mind, accustomed to weekly reports, becomes less agitated and alarmed at the sound of death. Medical men, informed of the endemic diseases of our climate, can more effectually devise an antidote." Pintard hoped that comparable figures on births and marriages would soon become available as well, for "the folio of *profit and loss*, in the records of human existence, cannot be fairly balanced without these tables."[33] Taken together, as some of his contemporaries agreed, vital statistics had a varied potential as social barometers. With them, "The medical philanthropist will rejoice to see how the smallpox declines. The economical calculator will find data for estimating the probable duration of life. The friends to penitentiary establishments, will admire the fewness of capital executions."[34]

Whatever their interest for public officials, economists, and scholars, bills of mortality continued to hold only limited interest for physicians. True, most of the bills that were being published about 1800 were broken down according to causes of death, and by then these lists generally included a considerably broader range of causes than their counterparts fifty years earlier. However, physicians could have little confidence in such documents as long as they knew that the bills were based upon information supplied by sextons, graveyard superintendents, and other nonmedical personnel. Such individuals' knowledge of medical terminology and their ability to distinguish between diseases clearly left nearly as much to be desired in 1800 as in the 1660s, when John Graunt was attempting to analyze the London bills of mortality.

When physicians themselves supplied information for local bills of mortality, it was not unreasonable to expect that they would help improve the content and accuracy of the bills. In practice, however, even the doctors could not be relied upon. Pintard, for one, found that although the mortality totals for New York City during his time were probably accurate, physicians were frequently careless or uninformed in assigning the causes of death. As long as this situation prevailed it was useless to think of obtaining a "correct judgment of the diseases which destroy life" from American bills of mortality.[35] In fact, it was not until the 1840s that physicians began to devote themselves seriously to reforming the nomenclature for diseases listed in public mortality reports. Meanwhile, some impetus toward better definition of medical terms and concepts could be discerned in the enthusiasm for scientific classifications and nosologies.

A Rage for Classification

Classification was one of the late eighteenth century's most popular means of organizing human knowledge. The fast-growing fields of natural history—mineralogy, zoology, chemistry, and botany—all of them fields with critical problems in keeping track of large amounts of new knowledge, benefited greatly from new classification systems.

Scholars in several of these sciences adopted the pioneering classification of the Swede Carolus Linnaeus. American intellectuals became enthusiasts of Linnaeus' "natural" arrangement and nomenclature almost as soon as their European counterparts did. His conception of a universe organized and run according to precise laws of the Creator struck a particularly responsive chord in America. By the last quarter of the century, virtually all of the botanists exploring, collecting, describing, and cataloging plant specimens as part of the American inventory had adopted Linnaeus' or some other system.

Classification proved to be equally essential for the various treatises on materia medica being produced by the nation's physician-botanists. In Charleston, John Shecut, whose middle name was Linnaeus, understandably based his medical botany of the flora of South Carolina firmly upon the great Swede's classification. Benjamin Smith Barton of Philadelphia, on the other hand, used a simple arrangement adapted from that of the British physician William Cullen.[36] Some physician-naturalists favored the drug descriptions, arrangement, and classification of Erasmus Darwin, whose *Zoonomia* had several American editions in the decade or so following its appearance in London in 1794. A few American writers even drew up their own classifications. Nathaniel Chapman, for one, arranged medicines in two large groups: local stimulants, with nine subgroups; and general stimulants, with three.[37]

Although individual preferences for the published classifications varied, early nineteenth-century physicians generally agreed that an orderly arrangement and standard terminology of the materia medica was essential. Such unanimity of opinion helped to pave the way for the eventual adoption in 1820 of a national pharmacopoeia and to establish the tradition of revising it at ten-year intervals.[38] That large measure of accord, however, contrasted sharply with a deep division within the profession over the value of medical classifications per se.

Reasonably well-read American physicians about 1800 were familiar with a variety of arrangements of diseases, classifications that had been constructed during the previous half-century or so by such European scholars as Linnaeus, François Boissier de Sauvages, Cullen, Erasmus Darwin, and Philippe Pinel. These classifications, called nosologies, served the medical profession well, at least so far as they fulfilled their promise of organizing the facts of medical knowledge.[39] Unfortunately, however, some were used as little more than frameworks for the presentation of the eighteenth century's highly theoretical disease concepts in some formalistic system or other. Such classifications quickly lost credibility among those who prized observation over hypothesis. Turn-of-the-century Baconians accordingly bitterly criticized nosologists who seemed to have forgotten their practical roles as straightforward organizers of medical knowledge.[40] At the same time, the belief in the existence of numerous specific diseases which was implicit in the nosologies (in opposition to a single generalized state of illness) came under vigorous attack from some late eighteenth-century medical systematists, who saw it both as an unproved hypothesis and as a threat to their systems.

The simple system of the British medical theorist John Brown was especially popular in the United States during the three or four decades after 1780 because of its frank challenge to the idea that disease entities were as distinct, or their study as complicated as the nosologists implied. "Nosologia delenda," Brown demanded—nosology must be destroyed.[41]

This message was subsequently carried to the American medical profession even more forcibly by its own systematist, Benjamin Rush. Like Brown, Rush adopted a hypothesis of the fundamental unity of fever, an attractive theory under which, as Boston's John Warren remarked, "the difficult science of medicine was . . . reduced to an easy art." Rush came to view the artificial division of ailments into genera and species as "the monster nosology," a medical heresy to be fought and eradicated. For a quarter of a century he passed this belief on to hundreds of students at the University of Pennsylvania. This group, together with numerous other disciples, became an implacable hard core of opposition to the entire subsequent antebellum effort to study diseases as specific entities.[42]

Despite this emotional opposition, nosology continued to have its followers. A number of disease classifications were planned by Americans, and some were launched. Edward Miller of New York proposed but did not actually complete a brief classification and nomenclature of some of the epidemic diseases, arranged by their remote causes. A few years later Franklin Bache presented preliminary ideas for a nosology to the College of Physicians of Philadelphia, although he never elaborated on them. In 1812, however, John B. Davidge of Baltimore constructed a full-scale nosology, and in 1818 David Hosack of New York devised another.[43]

Davidge, who divided diseases into four classes, seemed dazzled by the prospect of achieving medical order through literary arrangement. "The science of disease," he wrote, "is a grand whole, and like every other science, is made up of parts. The first act of nosology, is to separate these parts, and arrange them into order and system, according to their approximations in character."[44]

Hosack's system included eight classes, subdivided into orders, genera, and species, and organized on the basis of diseases which required the same general kinds of treatment. Hosack agreed with Linnaeus that "systematic arrangement . . . is the Ariadnean thread, without which all is confusion." He saw, however, that nosologies could not be constructed as botanical classifications were. His intention, therefore, was more "to convey a distinct enumeration of the character or pathognomonic symptoms of diseases, and to form those associations which are connected with their cure, than to observe the rigid rules exacted by the naturalist in the formation of genera and species."[45]

Early nineteenth-century physicians found a variety of uses for nosologies. Medical school professors, for their part, saw them as helpful aids to sorting out medical knowledge in the classroom: "By the rules of nosology, or systematic classification of diseases, the student very soon learns to distinguish diseases of a different nature from one another, and which, consequently require remedies not merely of a different grade in their qualities, but of a different kind."[46] Some sort of systematic arrangement, as in a nosology, was also recognized as being essential to the effective practical examination of disease in a community. In 1811 William Currie used the nosology of William Cullen as a framework for presenting the data he had collected in his survey of common American diseases.[47] Subsequently, other nineteenth-century medical authors used nosologies to help them organize and cope with increasingly complex questions of disease etiology, diagnosis, and pathology.

In fact, even the most vigorous opponents of nosology found that they could not entirely dispense with it. Rush modified his own unitary theory of disease by dividing fevers into three broad classes, each incorporating certain "states of fever." He went on to subdivide his Class I into sixteen states, Class II into eleven states, and Class III into eleven more, an acceptance of nosology which might well have given his teacher William Cullen satisfaction. As one of Rush's followers eventually conceded: "Some classification of diseases is required by practical convenience."[48]

Horatio G. Jameson in 1817 thought of nosology as "only useful as a dictionary," but in the chaos of early nineteenth-century medical terminology, this by itself was no mean contribution. Samuel Latham Mitchill had noted a few years earlier that "physicians and others generally choose to call diseases by such names as they think preferable, without reference to books," and he urged readers to put up with these idiosyncrasies for the time being.[49]

The medical classifications and their terminologies were, of course, arbitrary expedients which rarely completely satisfied anyone. However, within their limitations they helped clarify the medical ideas and problems of the antebellum period. It became evident that nosologies offered increased precision of both meaning and usage in medical terminology. The nosology of the English physician John Mason Good, published in 1817, was widely acclaimed in this country for its simplified terms and its contribution "to the rational reformation of the nomenclature of nosology," and at least one reviewer went on to criticize Hosack's nosology specifically because of its barbarisms of language and nomenclature, uses which seemed to represent a "step backwards." As it happened, neither Good, nor the organized medical profession, nor a new national pharmacopoeia, nor even the *North American Review*, which was enlisted in the problem in 1822, could do much to improve American medical language in these early decades of the century.[50] However, when a discernible measure of reform in terminology did take place, beginning in the 1840s, it occurred largely within the framework of practical nosologies. Classifications thus took on even more significance as adjuncts of the quantitative approach to medical and health concerns. For medical investigators, for editors and teachers, for health officials, and for reformers arranging their statistical lists of the causes of death, they were keys to increased precision and certainty.

Some of the findings of the health inventory, contributing as they did both to local and to national pride, provided a basis for an optimistic view of America's future. Other findings, of course, were far more sobering. Large parts of the inventory reflected a decidedly dark side of society, one which would require much attention if the health of the nation's citizens was to be maintained and improved.

2

A Calculus of Medical Philanthropy

The national inventory was undertaken, in part, to shed light upon a cluster of social problems existing after the Revolution, several of which bore directly or indirectly upon health. As it happened, in Europe the perfectionism of the Enlightenment had begun to spawn practical solutions for some of the problems. In the process, an impressive roster of reformers emerged. Among them, John Howard worked for improved prisons and hospitals, William Carey for foreign missions, Valentin Haüy for the education of the blind, Philippe Pinel for better treatment of the insane, William Wilberforce for the abolition of slavery, Count Rumford for organization of services for the poor. Most of these reformers made clear how necessary statistical information was, both in defining the nature and extent of given problems and in gaining attention and support for specific benevolent activities.

By and large, the Americans who introduced the various European reforms into the United States did not doubt their capacity to cope with the country's social problems. In fact, in the perfervid religious atmosphere of the Second Great Awakening, many people seemed to welcome such tasks as ways of testing their faith. The same people also welcomed the availability of statistical methodologies. Statistics permitted them to make a systematic approach to medical morality. It gave them means for measuring the steps being taken to make earthly society a better stopping-off place on the way to the final reward in heaven.

About 1800, the members of the American Protestant community were already involved in an intense and essentially quantitative mission, the spread of Christianity throughout the land. They sometimes couched their mission in language similar to that of Jeremy Bentham, for realization of their aims would indeed, they thought, ensure mankind's greatest possible happiness in this life and afterward. One concerned churchman felt that, given the unabated doubling of the nation's population, if care were not taken, the current generation of unbelievers would doubtless increase geometrically. "And surely, next to the sin of Adam will be ours, if the future millions of this new world should perish from our neglect."[1]

Communities moved to meet this religious and demographic challenge by building more churches, creating seminaries and colleges, establishing religious periodicals, and supporting a variety of religious and benevolent institutions. The first half of the nineteenth century saw a staggering proliferation of such bodies: Sunday schools, foreign and domestic missions, Bible and tract societies, orphans' homes, and charitable institutions of all kinds. Through scrupulous record-keeping, clerical leaders kept careful watch over the numerical expansion of these projects or institutions and made continued growth a measure of their own evangelistic dedication.[2]

This is not to say that religious groups were equally effective against the many social problems of the day. The inability to arrest the calamitous mortality and decline of the American Indian population in the late eighteenth and early nineteenth centuries was one of the more conspicuous failures of the Protestant establishment. And even more disastrous was its ineffective effort to promote the resettlement of any significant number of American blacks in African colonies.

The promotion and support of medical philanthropies, on the other hand, was a decided success, perhaps because such involvement focused upon projects which were closer to home and which benefited primarily the white population. The medical community was naturally heavily involved in such projects, and doctors became essential partners with clergymen and zealous laymen in creating, organizing, and managing health-related institutions or activities for the delinquent, the poor, and the otherwise disadvantaged. Most people tended to agree that the mending of bodies had a close connection to the saving of souls and promoting of morality. The physician was "in a fair way to add immense influence to the cause of religion . . . certainly much more than if he had been simply a preacher."[3] As the ranks of the medical profession grew in the nineteenth century, the physicians became increasingly conspicuous participants in society's attacks on poverty, prostitution, crime, and intemperance, as well as upon diseases themselves. As such, they were not only key figures in an emerging network of medical philanthropies but some of the most thorough statisticians of all kinds of social reforms.

The Rationale of Medical Benevolence

As administrators of medical benefactions, physicians shared many of the dilemmas and ambivalences of society in general, particularly with regard to the problem of poverty. Medical leaders continually pointed out that good physicians had a clear ethical responsibility to serve the poor. Some doctors were aware, however, that to be at all effective, medical philanthropy had to be part of a broader social benefaction aimed at reducing poverty. Benjamin Rush, for one, saw that poverty "exposes men to being soldiers and dying with yellow fever. It exposes women to diseases from hard labor, and to seduction, and both sexes to famine, strong drink to obviate hunger, and to the diseases from cold.[4] And Samuel Latham Mitchill made it clear that "unless the means of living shall be provided for those who are saved from the ravages of smallpox, they are but reserved, after the utmost sufferings of want and misery, to become, at last, the victims of some other disease."[5]

Early nineteenth-century Americans, however, were not at all clear or of one mind as to whether all of the poor were equally entitled to have their poverty relieved by society. Malthus had urged that public arrangements for the poor, the sick, and the aged gradually be done away with, and provisions of the poor laws in Great Britain hardened perceptibly as the century went on. The American Minister in London, Alexander H. Everett, found this phase of Malthus's doctrine "revolting" and perceived that "it leads of necessity to a low estimate of human life."[6] James Fenimore Cooper expressed the belief that Americans in general were too humane to countenance such a doctrine. He also felt that American physicians were more benevolent toward the poor than their Old World colleagues were. "Human life," he commented, "is something more valuable in America than in Europe, and I think a critical attention to patients more common here than in Europe, especially when the sufferer belongs to an inferior condition in life."[7]

Whatever the truth in that opinion, many Americans still harbored a deep doubt about the very morality of being poor. As in an earlier day, no one minded providing relief to those belonging to the same class or community, those who were churchgoers and generally respectable. Ever since the American Revolution, though, society seemed to have been acquiring increasing numbers of poor people who were unknown or otherwise unwanted. These included many foreigners, but also native prostitutes, criminals, drifters, fortune hunters, free blacks, and Indians—individuals who were rarely self-supporting for long, who all too often eventually became confirmed paupers, and who sooner or later required medical care.

The perceived demands of these groups for charity was a growing and annoying burden on the rest of society. Hoping to reduce this burden, community leaders went to considerable pains to measure it and to determine just who was eligible to receive relief. Assessing medical needs involved not only evaluating people's diseases or physical condition but attempting to measure their moral qualifications. The intention was to restrict benefits to the "deserving poor," denying them to the demonstrably dissolute, idle, and intemperate, the so-called professional paupers who fell outside the pale of respectability. There was great hope, of course, to turn many of those paupers into respectable and deserving citizens, through evangelism and moral suasion, but meanwhile, they remained a disquieting problem, a vast sore festering in society.

By the 1830s medical spokesmen were warning their colleagues about the unfortunate consequences of indiscriminate charity, particularly the threat of chronic pauperism, which might become as widespread as it was in England. One physician suggested that it was "the duty of every good citizen, to lend his influence towards discouraging the extension of this system; it is our duty professionally, to do this by curtailing the measure of gratuitous aid, and insisting upon remuneration proportionate to the patient's ability.[8] The question of medical charities had been thoroughly discussed in Great Britain, and a conservative position had been taken by most of the British medical establishment. Many American physicians and moralists agreed with that position. But difficult questions remained. How was society to identify the deserving poor? Who was to be given free medical treatment? Who was to be rejected for charity? Should the profligacy of undeserving paupers be rewarded as much as the virtue of the respectable poor? Could respectable society permanently ignore the health needs of the unrespectable?

No clear-cut answers to such questions existed, of course. Certificates of good character were required for admission to some medical institutions but not to others. Even where such regulations existed, however, the needs of an individual patient often led to the bending of the rules. As a Boston dispensary physician put it, "Where uncertainty exists charity induces the attendant to give his patient as favorable a character as possible."[9]

The distinction between "deserving" and "undeserving" poor remained a subjective one. How and whether to apply this distinction to the incidence of urban disease and mortality was debated by moralists, public officials, and physicians throughout the century. Organizers of every local medical-care institution had to come to grips with this distinction, to study the local data on poverty as a means of reaching decisions about which elements of society the institutions were to serve. In every community the alliance of physicians, religious leaders, and other philanthropists thus performed a quasi-moralistic, quasi-scientific operation. Health decisions were made from a set of preconceived notions and then enforced by statistical knowledge. The data were used to mold public opinion, raise money, and obtain the legislation necessary to build the institutions for the respective communities. In the process, people came to a better realization that maintaining the public health was an ongoing responsibility, not one satisfied by spasmodic public reactions to occasional epidemics.

Quantitative information about the unwanted elements of society was never very complete, but some medically significant data about them could usually be found. In 1803 a New York City police magistrate estimated for the *Medical Repository* that there were then some 1,050 known female prostitutes and 263 male pimps in the city, plus about the same number of "demi-reps, kept mistresses and private concubines"—in all, one-thirtieth of the population. The editors noted that, since the prostitutes were "under no regulation of the Health Office . . . the venereal disease is kept constantly in action, and diffused far and wide among both sexes." As the nineteenth century went forward, the statistics of prostitution became a matter of progressively greater concern among public-health personnel.[10] However, for much of the period the data were used much more conspicuously to fuel the moralists' campaigns.[11]

Destitute alcoholics were notorious for their habit of moving frequently from one town to another. A Boston physician recorded that "intemperate men are more transient than sober men. They more rarely own real estate, or have fixed businesses, and therefore change their places of residence more frequently than other men." A corollary was that a given drunkard would show up on the books of physicians or charitable institutions in almost every place to which he wandered.[12]

Physicians and other civic leaders were aware that, more than any other factor, it was the urgent need of these and other rootless and indigent classes that was forcing cities to build many of their new facilities for medical care.[13] Taxpayers naturally complained of the mounting costs. At the same time, health officials and editors complained that the rates of death and disease for these classes were so high that the normal statistics were inflated and gave their communities a bad image. As early as 1805, New Yorkers were attributing the excessive numbers of deaths listed on their bills of mortality specifically to destitute immigrants and transients.[14]

Residents of other cities made similar observations, and, as often as not, attributed the excess disease among such classes to deficiencies in morals. Charles Caldwell observed of New Orleans in about 1820 that "while the careless and the licentious suffer, the temperate, circumspect, and orderly portion of the community usually escape." Although more precautions against disease were needed in that warm climate than in the north, they were only what was dictated by good sense and moral living. It was doubtless too much, he felt, to expect sailors and laborers to reform. "But could the strangers of better education and higher standing, who, on business, or for other purposes, repair to New Orleans be induced to govern themselves by the same rules of discretion and propriety, which mark the conduct of the settled inhabitants, both their health and morals would be much more secure." Caldwell's subsequent public sanitary program for New Orleans was closely linked to his suggestions for personal hygiene and moral behavior. He considered public health activities useless without a thorough medical and moral reform of every individual. Caldwell had no doubt that "the disease and mortality which affect so materially the character and interests of the city, are more attributable to the faults of the sufferers, than to those of the place."[15] His conclusion was hardly novel. For all those swept up by the religious perfectionism and evangelism of the day, it eventually furnished a dynamic starting point for the organized sanitary reform in Philadelphia, New York, Baltimore, Boston, and other cities, as well as in New Orleans.

The Growth of Medical-Care Institutions

Every community had to determine for itself what its provisions for the sick poor should be. Many smaller towns or parishes long continued the eighteenth-century plan of appointing or contracting with physicians to look after the ailments of the destitute. However, as cities grew in wealth and size, they required more extensive formal arrangements. Through private benevolence and governmental provision a variety of such arrangements gradually came into existence. Not only moralistic, but demographic and other practical considerations stimulated interest in forming such institutions. It was clear to some community leaders that "the number of afflicted poor will increase in a compound ratio to the increase of inhabitants." Accordingly, society had an obligation to assist those individuals medically, "not only because it is commended by heaven and imposed on our feelings by nature," but also to prevent the poor from turning violently against the rest of society.[16]

The imposing structures in which the new institutions providing medical care—almshouses, hospitals, infirmaries, asylums, and others—were sometimes housed were tangible symbols both of the community's philanthropy and of its well-being.[17] At the same time, the patients in these institutions served as yardsticks for measuring the disease and mortality in their respective communities. Such data became useful supplements to or substitutes for other local sources of information on health, such as the bills of mortality.

Administrators of the medical-care institutions became data collectors and compilers out of practical necessity. Initially, substantial demographic evidence of potential patient populations had to be gathered to persuade legislators or private contributors to lend their support. Once the institution came into being, both its continuation and any plans for expansion had to be justified by extensive statistical data, often compiled in regular reports. Finally, money from both public and private contributions had to be accounted for, usually in the same reports.

In most communities, almshouses, or poorhouses, were the first kinds of institutions providing medical care to be founded. Although only Philadelphia, Boston, New York, and a few other towns were large enough to support such institutions during the colonial period, many others erected them in the nineteenth century. In fact, throughout the antebellum period, there were more almshouses than any other kind of medical-care institution, far more than general hospitals. Almshouses served as shelters for the aged, the homeless, and other destitute persons. Since there was a great deal of illness among the inmates, another significant function of the poorhouse was as a hospital. Almshouses often provided separate wards for the sick and had staffs of attending or consulting physicians.

Medically, the almshouses were often thought of as "*hu-manity's commons*, where the useless and incurable are 'turned out to die.' " A sampling in 1830 of the inmates of the Bellevue almshouse characterized them as follows:

> With a few accidental exceptions, they consisted of persons be-longing to the most indigent class of society, labourers, hired serv-ants, etc. Many of them had been long addicted to intemperance in drinking, and, among the females, a considerable number had been given up to the habits of prostitution. Of the number received only a small proportion were natives or residents of New York. The principal part were emigrants . . . most of whom had recently arrived in this country. A large number consisted of strangers from different parts of the United States, who were either casually here, or had very lately obtained a settlement.

About one in nine of the total were identified as blacks or mulattoes.[18]

The bare wards of the almshouses afforded few if any com-forts to chronically ill residents, but because they were the repositories of a variety of ailments, they provided clinical experience for many city physicians and a much-needed clin-ical setting for the instruction of medical students. Where de-cent records were kept, moreover, these institutions provided some of the earliest substantial data for American clinical re-search.[19]

The hospital, of course, was usually much better organized than the almshouse to provide medical care. However, in 1800 there were even fewer of them than almshouses, and for sev-eral decades of the new century, despite the best efforts of medical leaders, surprisingly few hospitals were built. Before 1800 hardly a half-dozen medical care institutions—most of them hospitals or asylums—had been established in the United States. A slightly larger number emerged between 1801 and 1825, but in the following thirty-five years, scores of new med-ical institutions sprang up.[20] Some were designed for special groups of patients; they included a cluster of marine hospitals, insane asylums, and scattered small hospitals for blacks. The remaining handful, which were chiefly in the largest cities, served more general populations. The earliest hospitals in

Philadelphia, New York, and Boston were the products of Protestant civic philanthropies. However, New Orleans's Charity Hospital was originally a Catholic institution and, as the nineteenth century progressed, the first hospitals in Saint Louis, Buffalo, and several other cities were those organized by Catholic groups. Although all of the hospitals had their defects, sometimes serious ones, they were a decided improvement over almshouses in providing medical care, and as such they satisfied a demonstrable need.

Though hospitals were not usually restricted to the poor, they were nonetheless founded with primarily the poor in mind. New York City physicians around the turn of the century, in fact, thought of their recently opened hospital as a "House of Charity," but it was not intended to compete with the almshouses, which were provided for the confirmed paupers, particularly those "labouring under incurable decrepitude and continued ailments of any kind." Hospitals, on the other hand, were designed to serve the respectable poor, to treat their acute and curable conditions, mental diseases, and obstetrical cases. A major rationale for building hospitals was to save the respectable poor from having to go to the almshouses and be stigmatized as paupers.[21]

Certainly there was some admixture of the two kinds of poor in both kinds of institution. When beds were available, sick paupers were occasionally admitted to the public charity wards of hospitals, as well as to the almshouses. Normally, however, in the early decades of the century character references or morality tests were standard requirements for admission to the hospitals and to other medical-care institutions except for almshouses.[22] Once charity patients were admitted, of course, records of the individuals were needed to ensure that the hospital received reimbursement from the public authorities.

Although most hospitals sought to take in paying patients along with charity patients, and usually had special facilities for them, the relative numbers of each varied considerably from year to year. The New York Hospital in 1813 treated 479 paying patients along with 857 nonpaying patients. During its first seventy-five years or so, the Pennsylvania Hospital struck an almost even balance of paying and charity patients, though in 1829 the proportion of charity patients increased markedly: 561 paying to 782 charity patients. The Massachusetts General Hospital often had sizable majorities of reasonably well-off patients; in 1836, for example, it admitted 291 paying patients and only 194 charity patients, and in 1840, 194 paying and 168 charity.[23]

Ideally, hospital benefactors would have liked most of their contributions used for the care of native sick people, but few hospitals were in a position to ensure this for very long. In 1813 the patients treated in the New York Hospital were almost equally divided between natives and the foreign-born. However, as the century went on and immigration accelerated, to achieve even a rough balance of the two segments of the population became progressively less realistic. New Orleans by midcentury provided an extreme example of the marked shift of its hospital population to foreigners. Samuel Cartwright observed in the early 1850s that the great Charity Hospital had changed in complexion to a point where hardly 150 patients a year were natives of Louisiana, "whereas the various wretched governments of Europe crowd its wards with from twelve to fifteen thousand patients annually." In New Orleans, as in other antebellum cities south and north, the dismaying prospect of having indefinitely to take care of Europe's multitudes of sick paupers may well have turned many a potential donor away from hospital philanthropies to restrictive nativist politics. At least some donors undoubtedly agreed with Cartwright that "hospitals are nothing but anti-republican sores on the body politic."[24]

However such feelings about foreigners may have affected the rise and maintenance of hospitals, Americans were still sufficiently carried along on the wave of religious benevolence to give rather extensive support to several other types of institutions that provided at least some medical services. Urban dispensaries alone grew from four in 1800 to about sixty by the time of the Civil War. Offering free consultations, prescriptions, and home visits, the dispensaries formed a source of "outdoor" medical relief intended to keep the poor out of other institutions. In some cities, as in New York, the dispensaries eventually took over the free public vaccination programs for the poor, and throughout the century they provided their young staff physicians with firsthand awareness of social and public health problems, as well as with extensive medical experience.

Doctors were also in ever-greater demand in prisons and orphanages. Others were required to staff the increasing numbers of infirmaries and clinics that arose to deal with eye and ear diseases, lung ailments, obstetrics, and eventually other medical conditions. And even greater were the demands of the asylum hospitals, which sprang up around the country in response to the growing numbers of the insane, the blind, the deaf, and the mute.[25]

Deaf-Mutes and the Blind

Building institutions for the benefit of the deaf, mute, and blind was one of the most popular medical charities from the 1820s onward. In fact, the desire to know how many of these unfortunates there were in American communities was great enough to count them in the federal census beginning in 1830. Philanthropists were drawn to these groups because there was real reason to think that they could be helped. To be sure, medical science had not yet done much to ameliorate their maladies, but educational innovations of dedicated European teachers were achieving remarkable new levels of communication for the deaf and blind. Charities for these people became exciting and hopeful ventures. One encouraging realization was that, since most of the afflicted came from the more indigent classes of the community, there was a real possibility of reducing the community's burden of permanent charity. Through education, it now appeared possible to give the afflicted "the means of earning their own livelihood" and by so doing "to take from society so many dead weights."[26]

The first institutions of these types in the United States thus quickly led to others. The founding in 1817 of the Connecticut Asylum for the Education and Instruction of Deaf and Dumb Persons was a direct product of New England religious zeal. The Congregational clergy conducted a survey which indicated the presence of a substantial population of deaf-mutes in New England, and subsequently the ministers raised funds for the asylum.[27] This asylum became a model for five other institutions for the deaf and mute, founded by the 1830s in New York City, Canajoharie (New York), Philadelphia, Columbus (Ohio), and Danville (Kentucky); within a few more years others sprang up. Similar establishments for the blind were also founded in the early 1830s, followed by a steady number of additional asylums over the years.

Like the founders of hospitals, community leaders who proposed institutions for the deaf and blind took considerable pains to gauge local needs in terms of numbers of potential users. And, like their French and English counterparts, they quickly found that careful record-keeping and detailed statistical reporting were essential in running their institutions. Legislatures, of course, demanded financial reports on the blind or deaf students whose tuition was supported by tax money. Tract societies consumed endless data about and testimonials from the blind or deaf students who received free Bibles or other literature.[28] There was also a need for statistical data of all kinds to use as ammunition in soliciting money from the wealthy for new buildings, improved facilities, and larger staffs, as well as for attempts to extend the benefits of their particular philanthropies to yet-unreached unfortunates.

Lawmakers were eager to obtain some overall idea of how many deaf-mutes and blind there were and of how many asylums might be needed for them. European experience indicated that approximately one person in every 2,000 of the population was likely to be deaf and mute, and this figure furnished a convenient, if rough, basis for discussion. However, censuses taken in Vermont, New York, Pennsylvania, and other states in the decade or two after 1818 indicated some differences. Nationwide, the federal census of 1830 counted 6,106 deaf in the total population of 12,901,049, which gave a ratio very close to the European statistic.[29] Lewis Weld, Principal of the Pennsylvania Institution, figured in 1828 that six institutions should be adequate to educate the current number of deaf-mutes and to take care of forseeable population increases.[30] This thinking would have made it unnecessary for every state to erect such asylums, but this projection was quickly outdated. By 1851 fourteen states were supporting institutions for the deaf, and a similar number supported establishments for the blind.

Actually, few responsible persons were satisfied during the decades of the twenties and thirties that their information about the number of deaf and blind was very good. Most were certain that the various census enumerations, federal and local, failed to account for large numbers of the afflicted. T. Romeyn Beck of Albany noted in 1836 that the proportion of deaf-mutes in the population was increasing rapidly, but he confessed that "the facts on this point are not numerous." And he found nothing in available data to explain why there were considerable variations in the proportions of the deaf living in different states.[31]

About the same time, Thomas Gallaudet and the directors of the American Asylum for the Deaf at Hartford proposed enlarging their inquiries to obtain "correct statistical views, with regard to the actual condition . . . of the deaf and dumb." They proposed that the clergy and medical men of each state should together conduct annual investigations. From these, they thought, data could be compiled which would not only help evaluate the extent and effectiveness of benevolent efforts but also shed light on the medical aspects of deafness. Several matters seemed particularly important:

> the sex and age of the individual; whether deafness is owing to some original, constitutional defect, or was produced by disease or accident, and, if so, in what way, and at what time; whether there are other cases of deafness in the same family, or among any of the ancestors . . . ; whether the deafness is total or partial; whether any medical means have been employed to remove it, and the results of such efforts; what are the circumstances of the parents or friends of the individual; . . . whether the individual has been taught any mechanical art or trade, or is engaged in any regular occupation.[32]

Detailed statistical inquiries into blindness were also needed. Dr. Samuel Gridley Howe and the trustees of Boston's Perkins Institution agreed in 1833 that their fellow citizens would be startled to know the exact numbers of their unfortunate blind neighbors who "sit and wile [sic] their long night of life away, within doors, unseen and unknown by the world." The proportions were not presumed to be the same in every town or country, because of varying soils and climates. However, for any given country, the frequency of the malady seemed to be a matter of inexorable, but computable, probabilities, once the basic statistical information was available. "It will be found, that the proportion is at all times about the same, in the same countries; for not only is the proportion of those who shall be born blind, decreed in the statutes of the Governor of the world, but the number of those becoming so, by what we call accident, is regulated by laws as infallible and invariable; and it is as little probable that by any accident, all mankind should lose their eyes, as that by any precaution all should preserve them."[33]

On the basis of data from the 1830 census, Samuel Hazard, a Philadelphia editor, hypothesized that proximity to sea air was a determining factor in the incidence of blindness: states furthest from the seacoast or closest to large freshwater lakes had the least amount.[34] Daniel Drake, Cincinnati's leading medical man, agreed that such findings might help to pinpoint the responsibilities of local communities to provide for this "melancholy aggregate of blind persons." Needed were not only institutions for the education of the blind, he thought, but also compulsory public provisions for treatment of eye diseases, to prevent potential blindness from developing, and its victims from depending on public charity: "Without the spirited co-operation of society, the profession cannot accomplish all which should be done. Endowed Infirmaries should be instituted in all our principal towns, where those who labor under chronic maladies of the eye, may be assembled, and placed in the circumstances, without which they are seldom cured. It is almost as necessary to take away the liberty of this class of patients, as if they were lunatics."[35]

The statistical data accumulating in the early eye institutions tended to confirm what Drake suspected. In Boston, a committee of clergymen in 1828 carefully examined records of the Massachusetts Charitable Eye and Ear Infirmary for its first four years. These data revealed a number of things: "that diseases of the eyes are eminently the diseases of the poor; . . . that eye infirmaries are the only places where the poor will apply, for relief of that organ; . . . that the amount of disease of these organs is vastly greater than would have been anticipated"; that, contrary to previous medical impressions, as many as "seven-eighths of the cases . . . have been cured or relieved"; and "that the amount of benefit conferred is incalculably greater than the amount of means expended," quite apart from the reduction of city expenditures for the poor.[36]

Neither Drake nor the Boston clergy thought of the increasing numbers and kinds of medical institutions as instruments for exerting any specific kind of social control over the poor, and probably few other American social or medical leaders did either. Basically, their endorsement of such institutions was shaped by the dictates of simple humanitarianism and medical responsibility. The new asylums, hospitals, and infirmaries of early nineteenth-century America provided central gathering points where the medical profession could reach those large numbers of impoverished sick people who would otherwise be neglected. Whether many of the institutions also provided an environment in which the "respectable" poor could retain their dignity is difficult to determine.

Intemperance

One of the severest threats, not only to this precarious respectability of the poor but to the health and well-being of all classes of society, was the excessive consumption of alcohol. In fact, few aspects of the new humanitarian impulse stressed the close relation of morality and medicine as strongly as the campaign against intemperance. Few causes did more to produce solidarity among early nineteenth-century philanthropists, and no social problem brought the medical profession closer to the many lay reformers. Likewise, no reform movement produced a greater volume of statistics.

The religious establishment, sparked by such clerical leaders as Jedidiah Morse and Lyman Beecher, launched a temperance crusade during the War of 1812 which by 1826 had created a national body, The American Society for the Promotion of Temperance, and by 1834 some five thousand local temperance groups were closely associated with churches. Well before 1812 members of the medical profession were taking their own substantial steps to combat drinking (by no means a new phenomenon in this country), and by the 1830s hygiene reformers made temperance a central objective in the improvement of individual health. All groups saw intemperance as a threat not only to the hoped-for expansion of the Protestant religion but to the prosperity of the nation itself.[37]

To the lay philanthropist, drinking was a vice which seemed to be spawning ever-greater poverty, crime, and degeneracy. To the humane physician, it was a habit "by which a considerable portion of the hard earnings of the poor, instead of being employed to provide wholesome nourishment, is appropriated to the purchase of a destructive poison."[38] Beecher noted the righteous indignation which the benevolent in every community felt toward those "degraded" individuals who "drink up their daily earnings and bless God for the poor house." The temperance movement was fueled as much by the anger of those who resented supporting such individuals with public tax money as by those who were concerned with the souls of the offenders.[39]

During the early decades of the nineteenth century, Americans consumed more alcoholic beverages per capita than ever before or since. Although the exact magnitude was not known precisely, contemporary estimates of consumption reflected people's awareness of the seriousness of the matter. One gauge of general consumption, of course, was the number of taverns in a given city. Benjamin Rush felt in 1809 that Philadelphia's taverns were not frequented as much as they had been forty-five years earlier, but there was now "a good deal of secret drinking" among all classes.[40] More informative was an inquiry made in New York in 1803 which reported that there were approximately twelve hundred taverns for the estimated population of sixty-four thousand: one tavern for every fifty-three people, including children. Only a year later, the figure had risen to fourteen hundred taverns.[41] Although not every city had quite so many drinking establishments, their proliferation, often in the shadow of the churches, was a matter of continuing concern to the reformers.

The American capacity for strong drink was truly something to wonder at. In 1818 the editors of the *Medical Repository* testified: "The quantity of ardent spirits drank [sic] by our people exceeds every thing of the kind, that the world can produce; the appetite for inebriating drink seems to be increasing and insatiable. The "dread of water," (a species of hydrophobia) is the wide spreading epidemic of the land. So few and small are the restraints to drunkenness imposed by the government, that the labourer swallows as much strong drink as he pleases. A quart a day is no uncommon quantity."[42] Some years later, medical observers reported that the consumption of three quarts of whiskey daily was not uncommon among soldiers along the frontier, "one quart . . . being required just 'to set them up before breakfast.' "[43]

The country's armed forces were special concerns of the temperance groups. Military service, of course, has always been accompanied by consumption of a great deal of alcohol, used sometimes to reinforce the will of forces in combat, often to help relieve the tedium of peacetime cruises or garrison duty. Nineteenth-century line officers were all too familiar with the problems of commanding soldiers who drank excessively, and medical officers coped with the medical problems.

As one such step, Surgeon William P. C. Barton conducted an experiment aimed at obtaining statistics on the medical effects of drinking by sailors. This experiment took place during the late 1820s on the USS *Brandywine*, which had a reputation as a "sickly ship," while it was cruising in tropical waters, much of the time near the West Indies. Barton organized a group of twenty midshipmen who volunteered to drink nothing but water during the entire cruise of three years. At the end of the voyage he reported that the group had compiled an unusually good health record. Except for one man, they had avoided all serious diseases and made rapid recoveries from the inevitable miscellaneous fevers, a record which he attributed directly to abstinence.[44]

Barton's findings did not, of course, bring about a "dry" Navy, even after he became Director of the Bureau of Medicine and Surgery in 1842. However, as part of the same spirit of reform that ultimately ousted prostitutes from the ships and abolished flogging, the findings did help bring about the abolition of the grog ration in 1862.[45]

The army's experience with drinking produced a far greater volume of statistics than did the navy's. Early data were so alarming, in fact, that they helped bring about the abolition of the army's whiskey ration in 1830. However, since whiskey was readily available elsewhere, the problem remained. Surgeon William Beaumont asserted in 1834 that intemperance at the army's Jefferson barracks in Missouri was the cause, "either directly or indirectly, of more than three-fourths of the diseases and injuries of this command." And another army physician, Samuel Forry, a few years later, concluded that Beaumont's observation was "equally applicable to every military station." The problem of drunkenness was clearly great enough to encourage commanders to cooperate with local reformers in establishing temperance societies at the garrisons.[46]

In the late 1830s Forry examined the army's medical experience with drinking with great care. It was evident to him that "up to the present day, the statistics of intemperance in reference to etiology, pathology, and therapeutics, have been so loose and unsatisfactory, as not to allow of any accurate deductions." However, the data of the accumulated army reports offered an extensive basis for statistical generalization. Forry thought that the relation of heavy drinking to the "causation of phthisis pulmonalis and epidemic cholera, has been abundantly pointed out in these statistics; and its intimate connection with febrile diseases, diarrhoea, dysentery and hepatitis, although not definitely determined, is yet so apparent that it is constantly dwelt upon in the reports of medical officers." He also found intemperance in the army closely associated with epilepsy, delirium tremens, meningitis, and with a wide variety of injuries and deaths from accidents. "What a long and frightful catalogue of ills!" he concluded. But he felt at least some satisfaction in knowing that the situation was even worse in the British army.[47]

Among other matters, the army's medical reports shed light upon the effects of climate upon excessive drinkers. Physicians in the South had long blamed intemperance, either separately or along with their hot climate, for much of the disease and mortality they saw, and one of them, David Ramsay, saw it as the clear enemy of longevity: "It makes young men old before they reach their thirtieth year, and brings them with all the infirmities and decrepitude of age to premature graves." Dr. Joshua E. White of Savannah felt that "rigid temperance" was one of the essential precautions against disease in that climate. And a health committee in Mobile blamed intemperance, which left people oblivious to any hygienic precautions, for much of the very high death rate in 1819 from yellow fever, particularly among transient laborers.[48] Forry's analysis seemed to confirm all of these impressions. So far as the reports from various garrisons indicated, inebriation itself occurred to about the same extent in the North and the South. But the reported deaths from drunkenness differed considerably: twenty-three per ten thousand troops annually in the South, and only two in the North. For virtually all of the diseases connected with intemperance, the death ratio was found to rise markedly in the hot areas of the country.[49]

The medical consequences of excessive drinking in the civilian population were delineated even more fully, because many groups and individuals contributed to amassing factual information. Among the myriads of sermons, reports, broadsides, and other publications on the subject, no single document did as much to alert people to the dangers of strong drink as Benjamin Rush's widely distributed *Inquiry into the Effects of Spirituous Liquors on the Human Body*. This small tract derived much of its persuasiveness from its "Moral and Physical Thermometer," a chart in which Rush ranked the various alcoholic beverages along a scale of the moral and physical degeneracy which presumedly resulted from their use (Figure 1). While it was in no way based upon experiments or statistics, this graph nevertheless gave readers an impression of having scientific authority. Whatever it lacked in precision it made up in its lurid catalog of fates held out for the morning drinker or the devotee of slings and pepper in rum. The tract established a vivid equation of disease with immorality for decades of temperance reformers to come.[50]

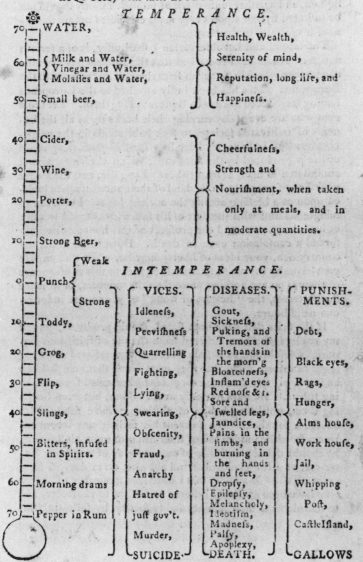

Figure 1. Benjamin Rush's "Moral and Physical Thermometer," from his pamphlet, *An Inquiry into the Effects of Spirituous Liquors on the Human Body* (Boston: Thomas & Andrews, 1790), p. 12. Courtesy of the National Library of Medicine.

More prosaic, but also more informative about the effects of alcohol, were the bills of mortality of various cities. These documents revealed steady increases in the number of deaths attributed to drunkenness during the early decades of the century. Boston's figures published in 1832 showed that only consumption, typhus, "lung fever," and dysentery had been regularly causing more deaths. New York was at least as bad, according to Dr. Charles A. Lee, though falsification of death certificates made the exact situation impossible to determine. Deaths from "drunkenness" in 1836 were substantially under-reported, he found, since that term was reserved only for "the lowest and most wretched class. If the relatives are respectable, or the deceased was a man of wealth, it is rare to find him included in this class. In such cases he will be more likely to come under 'sudden,' 'casualty,' 'apoplexy,' 'nervous,' etc."[51]

Temperance reformers, of course, used all of the statistical information they could find, so their polemics frequently had a distinctly quantitative quality. Lyman Beecher was among the clergymen who found it useful and effective to stress certain of these actuarial aspects of drunkenness in his sermons.[52]

Civilian physicians of this period attributed many general manifestations of poor health to excessive drinking. They also linked it to certain specific diseases, such as tuberculosis, gout, "liver trouble," dropsy, palsy, and tremors, but particularly to insanity. Rush concluded that, of the total number of insane patients at the Pennsylvania Hospital early in the century, one-third had gone mad through their use of strong drinks. A few decades later, Ohio officials and physicians were overjoyed to watch the percentage of alcohol-related insanity cases at the state lunatic asylum decline over an eight-year period from 14.5 to 3.5, and they attributed the decline to a revival of the temperance campaign in that state.[53]

Physicians in the South were persuaded that heavy drinking increased a person's susceptibility to yellow fever. It also led people to ignore the danger of the disease and to neglect personal hygiene. Some men, of course, resorted to liquor to drive away their fears of the disease. In any case, drinkers who did not die of yellow fever were frequently found to have died of delirium tremens.[54]

With the coming of cholera in 1832, many observers made an almost automatic correlation of the incidence of that dread disease with the consumption of alcohol, and particularly by immigrants. It was widely noted that intense fear of the epidemic by the immigrants and the poor generally "gave a great impulse to whiskey and brandy drinking," a serious setback to the temperance movement. An observer of the epidemic's progress in Montreal, however, suggested that temperance might be the best protection available against the disease. Despite the great mortality from cholera in that city, there were reported to have been almost no deaths among the members of its three temperance societies.[55]

In Albany, in the fall of 1832, the New York State Temperance Society looked into this matter much more systematically. It undertook an investigation of the backgrounds of all of the cholera victims of that city "with an express view to ascertain the degree in which the use of ardent spirits may be regarded as predisposing to the disease or disarming the system of its means of resisting it." Their findings, which were published both by the religious and the medical press, provided reformers with a statistical pattern of the drinking habits of 336 deceased adults, of whom 195 were natives and 141 foreign-born:

Intemperate	140
Free drinkers	55
Moderate drinkers, mostly habitual	131
Strictly temperate	5
Members of Temperance Society	2
Idiot	1
Unknown	2

Few other groups conducted such studies, but by then it was widely assumed that most cholera victims were consumers of alcohol. This belief continued at least through the 1850s, though by the time of later cholera epidemics intemperance was not considered as important a causative factor as housing, sanitation, and weather.[56]

The developing temperance movement found the medical profession playing a mixed role. During early decades of the century, physicians in every community prescribed alcohol so freely for their patients' ailments that its medicinal use contributed substantially to the overall drinking problem. Only in the 1830s did the medicinal use of alcohol begin to decline, chiefly because of temperance agitation. Meanwhile, of course, there were differences in physicians' personal use of alcohol. Not a few doctors became completely abstemious during this time, and a considerable number played vigorous roles as temperance lecturers and advocates in their communities. Some took the movement to the medical schools, a number of which formed their own temperance societies during the 1830s. As a result of such initiatives, an Eastern medical editor optimistically reported that "the reproach of intemperance can no longer be laid to the door of the physicians. They are everywhere the efficient instruments in the great moral revolution now going on."[57]

In some communities, the medical profession even earned praise for its "disinterested self-sacrifice" in campaigning against drinking. In so doing, it was thought that the physicians were voluntarily cutting off one of the most prolific sources of their own incomes.[58] One Boston practitioner, however, who identified himself only as J, was not entirely certain that much sacrifice was really involved. J examined the records of his own practice covering a period of three years and arranged his 663 patients in three groups. Of these, the "abstemious," who numbered 530, had 928 cases of sickness, or an average of 1.75 cases per individual. The "temperate" numbered 51 and had 135 cases of illness, or 2.66 cases per person. The third group, the "intemperate," numbered 82 and had 188 illnesses, or 2.29 per person. He noted that an upward correction should be made in the number of cases of the intemperate, whose "fickleness and migratory habits" often prevented necessary follow-up visits. J was unable to provide much data on the comparative length of sickness among the three classes, but he did produce figures for his charity patients which showed that "the intemperate constitute more than half of all the poor who have employed me." He also showed that, although intemperate patients who could pay brought him a considerable amount in fees, his loss in the uncollectible fees of the intemperate poor amounted to a still greater sum. There was already reason enough for "even the most selfish of our profession [to] join the temperance cause."[59]

Despite the amassed statistics, the reformers' exhortations, and the medical warnings, the early temperance movement was slow in showing any very substantial results. In fact, about 1830 drinking reached an apparent all-time high, with an estimated yearly consumption of $8\frac{1}{4}$ gallons of alcohol per "steady drinker." During the following decade, however, this situation changed dramatically, as the energies of ministers, housewives, health reformers, and others were finally harnessed and effectively organized in the movement. By 1840 the hygienist John Bell of Philadelphia, though conceding the defectiveness of statistics on the subject, estimated that the annual consumption of alcohol had declined as much as 57 percent from the 1830 figure to a mere $3\frac{1}{2}$ gallons per steady drinker. Somewhat more impressionistically, Daniel Drake in 1840 found "far less" drinking going on in the Ohio Valley than there had been twelve years earlier: less on steamboats and in taverns, less at political conventions and Fourth of July celebrations, and even less alcohol being provided by farmers to their hired hands to ease the extra labor of harvest time. Reformers in every part of the country noted the improvement that had occurred in society. Observers in Mississippi reported that the liquor shops or "dogerys," which in about 1830 had stood at every crossroad, had mostly been driven out by the mid-forties. In Massachusetts, moreover, Dr. John C. Warren figured that the temperance reform had increased the "physical strength" of the commonwealth by one-sixth; "the improvement in health, property, and happiness of the people is beyond calculation."[60]

Improvement was also noticeable in the drinking habits of professional people. By 1840 the widespread heavy drinking among clergymen that had scandalized Lyman Beecher and other early reformers had virtually disappeared. The same could not be claimed for lawyers and physicians. Still, even the local physician was less often characterized as "an excellent doctor if called when sober."

Perhaps the only place where drinking among physicians was systematically scrutinized at this time was Natchez, Mississippi. When Samuel A. Cartwright first settled in Natchez in 1823 along with two other new physicians, he found only one other practicing physician who did not drink, plus three more in nearby villages. "All the rest believed in the hygienic virtues of alcoholic drinks, and taught that doctrine by precept and example." Over the next ten years or so, he watched the arrival of sixty-two more new physicians, whose average age was twenty-five. Thirty-seven of them turned out to be temperate, and the rest were substantial drinkers. Later, when he followed up the careers of these individuals, Cartwright found in their longevity record a persuasive argument for temperance. In 1853, out of the twenty-five drinkers only three were still alive, while among the thirty-seven abstainers only nine had died.[61]

To be sure, gains in restraining alcohol consumption could not be assumed to be permanent. As Daniel Drake noted, the movement could lose its momentum quickly, "for all excitements, even political fermentations, religious revivals, and medical broils, are paroxysmal."[62] The increasing numbers of Irish and German immigrants, for their part, were actively resisting the attempts of native reformers to force unnatural habits of abstinence upon them. In the 1850s, to prove that plenty of work remained to be done, temperance advocates noted that alcoholism was still causing greater mortality in the United States than cholera, yellow fever, and smallpox combined. Nevertheless, that record was encouraging when compared with that of the previous generation.

The improvements achieved owed much to the effective organization and moralistic motivation of the temperance societies. The Parisian sanitarian Louis René Villermé wrote, " 'I consider the religious spirit of the United States as a powerful element in the success of temperance societies, which does not exist in our population.' "[63] He might have added that this moral spirit also helped make the members of those societies, as well as those connected with the new hospitals and dispensaries, exceptionally conscientious and energetic compilers of statistical information.

In its various manifestations, the orderly and quantitative character of medical philanthropy was both impressive and obvious. Just as the moral effects of benevolence could be measured by the total numbers of souls saved or of signers of abstinence pledges, so the medical effects could be calculated by such practical yardsticks as the numbers of admissions to almshouses or hospitals, numbers of dispensary physicians, and eventually by changes in the morbidity rates. Additionally, wherever medical-care institutions sprang up, they brought a new dimension to the medical world of their respective communities. They served as concrete evidence that the medical profession in early nineteenth-century America was becoming an organized, increasingly effective, and numerically impressive establishment.

3

Quantification in Orthodox Medicine

The physicians who took part in establishing and operating the various medical-care facilities were also looking for answers to a fundamental medical question: what kind of medicine should they endorse, both for those institutions and for use in their private practices? The attempt to answer this question led, fairly early in the century, to the tacit formulation of a consensus of orthodox medical ideas. This consensus led in turn to the emergence of a "regular" medical establishment—a growing cluster of mainstream medical societies, schools, and journals devoted to furthering these orthodox ideas and representing the large majority of American physicians—in contrast to numerous "irregular" practitioners and institutions which came on the scene later. As the century progressed, America's orthodox medical leaders, like those in Europe, were increasingly interested not so much in enlarging the current body of medical theory as in achieving a new high degree of certainty about medical knowledge. The expanded clinical studies of diseases and remedies which were thus undertaken were increasingly characterized by the systematic use of quantitative methods.

Medical Philosophy and Medical Numbers

It was one thing, of course, for certain learned men to suggest that American medicine could benefit from quantitative studies or from critical analysis of any kind. It was quite another matter for ordinary individual doctors without the same breadth of background to see that there might be a need for such study. It was also difficult for physicians steeped in certain aspects of conventional eighteenth-century medical wisdom to perceive the desirability of change, and least of all, change in well-established therapeutic practices. Like their European counterparts, American regular physicians in the early nineteenth century were far from agreement among themselves about several elements of their orthodoxy, not only as to the efficacy of specific therapies but as to the proper philosophical approach to treatment. Many practitioners were pragmatists who based at least some of their therapeutic measures on the results of accumulated empirical experience. However, others based their practices essentially upon the general formulas of one theoretical system or another. Allegiances to the differing philosophies had considerable bearing upon which doctor used numerical methods.

Many turn-of-the-century American physicians who pretended to formal learning still held to early eighteenth-century European theories of disease, those of Hermann Boerhaave, Georg Stahl, Sauvages, or Friedrich Hoffmann. Others swore by the more recent medical concepts of the Scottish physician William Cullen, while those who were still more up-to-date were adopting the system of Cullen's student, John Brown. A few individuals seem to have accepted the American Samuel Latham Mitchill's doctrine of *septon*, a chemical concept of disease and medicine. However, it was Benjamin Rush who formulated the body of theory that most fully represented the essence of medical orthodoxy in America during the first third of the nineteenth-century.

Rush's system drew heavily on the doctrine of John Brown. Greatly simplified, it assumed that there was really only one disease, a condition manifested by excess tension or excitability, to one degree or another, in the blood vessels. As a consequence, Rush maintained that only a single basic type of treatment was necessary: a generously applied combination of bloodletting and purging aimed at restoring the body's state of health. With this theory Rush accomplished what every American of his generation desperately wanted; he proved that Americans could, after all, equal the Europeans in scholarship, if only they would put their minds to it.[1] If his actual modes of therapy differed little in kind or substance from those of the Europeans, his concept of the vast *quantities* needed to produce desired effects was so extreme that Rush became famous in his own day, infamous to some.

From his influential, long-time position on the medical faculty of the University of Pennsylvania, Rush sent hundreds of new physicians out into the hinterlands of America indoctrinated in his simple therapy. These doctors, along with his other disciples, became therapeutic activists who considered it their duty to attack disease in the body as forcefully as possible, by using emetics in large amounts to ensure evacuations of bile, by using purgatives to obtain other vigorous evacuations of body wastes, and by repeated massive bloodletting, often to the point of fainting, to reduce inflammation.

The wide acceptance and extension of this "heroic therapy," as it soon came to be known, made the sale of drugs increasingly profitable. While many physicians continued to dispense drugs as part of their overall medical business, they began to encounter greater competition from independent apothecaries and druggists, and from general retailers who sold drugs along with other products. As the need for training people to compound the drugs grew, pharmacy began to become a separate profession, with its own schools, societies, and journals.[2] Meanwhile, wholesalers flooded the country with native and imported remedies of all sorts, creating a runaway commercialism in which the itinerant patent medicine vendor became as ubiquitous as the established apothecary.

Such developments called for a degree of control and precision that early nineteenth-century organized medicine proved unable to impose to any great extent. For one thing, the standard British and European pharmacopoeias were not entirely suitable for the New World. True, Rush's medical contemporaries created local pharmacopoeias in an effort to standardize the drugs they were dispensing, and the next generation established a national pharmacopoeia. At the same time, experimental chemists and pharmacists were familiar with the need to measure and weigh accurately the ingredients of their medicinal preparations. However, little of this sense of quantification or precision filtered down to those doctors who actually applied Rush's "heroic" therapy. Physicians drew blood by the ounce or dram, in amounts that differed for given diseases and according to the condition of the patient, but they were guided mostly by the individual patient's response to the treatment, and very little by the results of experiment or quantitative analysis. All too often they administered drugs with only vague rules of thumb for determining the size or amount of the dosage. The unitary theory of disease made treatment a simple, almost automatic process for Rush's followers. In their hands, "heroic therapy" produced readily visible physiological responses which often seemed to lead to cures. In fact, the treatment was so easy to administer, and the superficial effects so dramatic, that most doctors were for some time little inclined to question its value or subject it to any form of objective quantitative evaluation.

Baconian Concepts in Medical Thought

Rush's therapeutic system had its full share of critics, particularly scholars who found inadequacies in all formal medical systems. It became increasingly evident to such individuals that, whatever the apparent merits of the systems, they introduced a great deal of "looseness and irregularity" into medical reasoning.[3] Breaking away from this habitual vagueness and from the practice of medicine so largely based upon hypothesis thus became a primary concern of these critics.

The most promising turn-of-the-century alternative to having to rely on formal medical systems seemed to lie in expanding the search for the concrete facts of actual medical experience, that is, in adopting the empirical methodology of Francis Bacon. Actually, the Baconian ideal was something that almost all medical men at this time, including the most confirmed theorists, could endorse up to a certain point—at least the part of it that had to do with obtaining accurate information through observation and description. Many theorists, however, tended to use facts chiefly to justify their a priori systems. Their critics believed, with Bacon, that facts had to come first—from data gathering, measurement, experiment, and analysis—to provide a sound base for medical knowledge and practice.

Throughout Europe, "Baconianism" had long since become

synonymous with scientific method. Early nineteenth-century American clinicians received their exposure to its concepts chiefly through English and French medical writers. English medicine since the day of Thomas Sydenham had included a tradition of empirical observation and description that owed fully as much to Bacon as to Hippocrates. By the last third of the eighteenth century, in fact, an important nucleus of British physicians, including such men as John Gregory, Thomas Short, John Haygarth, and William Black, had come to regard the processes of original observation, empirical fact-finding, and numeration as essential in conducting clinical studies.[4]

French medicine had its own active tradition of empirical observation and data gathering. By 1800, a creative group of physicians had begun both to systematize these methods and to greatly extend the boundaries of medical observation through experiment. Administrators were now attempting to regularize, record, and analyze numerically every facet of hospital life and environment. At La Salpêtrière, for instance, Philippe Pinel carefully recorded and studied data on such matters as meals, hygienic practices, the influence of the seasons on the patients, and the amounts and actions of drugs administered, as well as the morbid condition of each patient. In another direction, French physicians were making the study of these conditions infinitely more precise by using new procedures and instruments. Xavier Bichat's painstaking analysis and classification of human tissues just before 1800 almost by itself opened up the new medical specialty that his generation called "pathological anatomy," and in so doing put a new value upon systematic postmortem examinations. Within a few years, moreover, such contributions as Jean-Nicolas Corvisart's rediscovery of percussion and René T. H. Laennec's invention of the stethoscope had immeasurably expanded the means of diagnosing disease in the living.

In the United States, a number of physicians and other learned men urged an end to reliance upon formal medical systems and proposed instead the introduction of up-to-date Baconian methods in the country's slowly multiplying institutions for medical care. The New York historian Samuel Miller thought that only these methods offered much promise for improving American medicine in the nineteenth century.[5] In 1807 Thomas Jefferson, whose antipathy to medical theorizing and appreciation for Bacon owed much to the French physician Pierre Jean Cabanis, urged the medical profession to reform itself by abandoning its reliance on hypotheses in favor of observation, experiment, and induction from "sober facts."[6] Similarly, the editors of the *Medical Repository* found in Baconianism the effective answer to "the absurd systems of ancient physicians [and] modern hypotheses scarcely less preposterous." It was all well and good for the Age of the Enlightenment to have its theories, they agreed, but not theories made out of whole cloth:

> Facts are the only rational basis of theory. Philosophers are no longer permitted to descend from generals to particulars, shaping them according to preconceived notions of their intimate relations, but are expected to proceed by a rigid examination and cautious assemblage of particulars to every general inference. This laborious process of reasoning, so favourable to truth, and so little flattering to indolence, to vanity, and to a creative fancy, requires the possession of an extensive mass of experiment, a various and judicious selection of facts, not only for him who would overthrow or construct a system, but for every one who would rightly exercise the art to which they belong.[7]

During the early decades of the century American medicine did start to absorb these methods. The practice of making postmortem examinations increased steadily, most conspicuously in the new almshouses and hospitals. By the mid-1820s conducting such examinations had become commonplace to physicians in Philadelphia and some other cities, and the first native textbook on pathology appeared in 1825.[8] Likewise in the 1820s, American clinicians began exploring the value of the stethoscope. However, since its use could be "learned only in hospitals, or where it can be applied to a large number of patients," the instrument did not spread very widely for another decade or more.[9]

Numerical factors also began to play some role in clinical investigation during this period. Well before 1820 physicians occasionally framed the reports of their experience with dropsy, diseased tonsils, syphilis, or some other condition at least partially in quantitative terms.[10] However, few if any of these doctors were at all systematic in gathering, organizing, or analyzing their data.

James Jackson of Boston at least had an idea of what numerical method required of the investigator. As early as 1805 Jackson learned about the method introduced by George Fordyce in England to organize case records by various categories and to arrange them in tables to expedite quantitative analysis. Jackson thought seriously of utilizing this procedure himself and even "had blanks printed for my cases, according to the plan of Fordyce. But the difficulties attending the plan in private practice discouraged me too soon." Jackson thus did not actually venture to undertake numerical clinical studies of his own until data began to accumulate at the Massachusetts General Hospital some decades later. Numerical analysis clearly was a demanding procedure, and Jackson could well understand why many of his countrymen preferred the simple medical systems of a Rush or a John Brown. "According to Brown," he wrote, "it is useless to record the phenomena of disease, to collate and compare cases. . . . The Brunonian system is popular, because it indulges indolence, as it exempts us from poring over the observations of others . . . and from the wearisome task of close observations, and slow and careful inductions."[11]

The difficulties Jackson experienced with numerical methods were indeed enough to daunt American clinical investigators for years. They also tended to confirm the suspicion of some that medicine could not attain the same degree of precision that other sciences enjoyed. J. Augustine Smith, for one, thought that only three of the four Baconian elements he considered necessary for achieving scientific certainty—observation, experiment, and induction—would ever be available to medicine. The fourth element, numerical calculation, he thought, was really "excluded from medicine: we must content ourselves, therefore, with the other three."[12]

A strong dissenter from this view was Samuel Jackson of Philadelphia. Jackson believed that the world of health and disease was indeed "susceptible of calculation, of being arranged into definite series . . . and may ultimately be reduced

to as much system and certainty . . . as the physical sciences."
During the mid-1820s he began to make plans to use some
such procedures in assessing his own clinical practice at the
Philadelphia Almshouse, particularly his success with the French
systematist François J. V. Broussais's "physiological therapy"
of localized bleeding.

Before such a study could be attempted, though, a means
had to be found to improve the Almshouse's procedures for
keeping medical records. Especially needed were case books
for each ward. Jackson obtained the books and then enlisted
medical students, particularly the residents, in the task of actually making and keeping the clinical records. This pioneer
arrangement made it possible in 1827 for Jackson to prepare
a limited report on his use of the Broussais therapy. As part
of the report he took pains to explain how the record keeping
worked: "The cases reported have been witnessed by numerous students; the records of the symptoms, of the attending
circumstances, and the treatment, have been kept by persons
(the students of the house) uninterested in the propagation
of particular doctrines, or plans of treatment, who are not
committed in their opinion, and who have no motive to ex-
aggerate the symptoms, to conceal errors, or otherwise to de-
viate from perfect exactness in the particulars they daily enter
in their journals.[13]

This initiative to provide systematic clerical arrangements
for recording clinical experiences was a significant development in American medicine. Jackson thought that calculations
resulting from such arrangements could eventually contribute
significantly to the search for medical certainty. As it happened, however, he did not actually attempt to analyze his
collected cases numerically or to compare the results of the
Broussais treatment with those of other treatments. Moreover,
he took little account of his unsuccessful cases. With a predilection for using his collected data to support preconceived
ideas, Samuel Jackson was not really the person who could go
very far with the numerical method. Much more promising
were those few American students who, in about 1830, were
beginning to obtain the necessary expertise in the hospitals of
Paris.

Louis and His Numerical Method

Following the Napoleonic Wars and the War of 1812, students from all over Europe, and Americans with them, were increasingly drawn to a Paris that had emerged as the acknowledged center of medical and scientific progress. Some wanted to study chemistry in the country of Antoine Lavoisier. Others went to absorb zoology with Georges Cuvier or botany with other teachers. Not a few got training in pharmacy, while others absorbed the new methods of teaching the blind and deaf.

Medical students flocked to the Parisian hospitals, where they looked for instruction to Baron Dominique-Jean Larrey, Guillaume Dupuytren, and other brilliant surgeons. Some sought training in gynecology, children's diseases, handling of the insane, or other emerging specialties. Still others hoped to learn pathological anatomy at its source and to get the advanced clinical instruction that was increasingly essential to an informed medical practice.[14]

Of Paris's many outstanding medical men, few attracted more foreign students during the thirties and forties than Pierre-Charles-Alexandre Louis. Louis had a somewhat unorthodox career, one in which all of the impulses to achieve precision on the part of the earlier French clinicians came to a focus. That career had started off unremarkably with several years of medical practice in Russia. Finding that unsatisfying, Louis returned to Paris in the early 1820s to immerse himself in the study and differentiation of disease. His self-organized project was conducted largely in the wards of La Charité hospital, in an arrangement facilitated by the director, his friend August-François Chomel. There he gave himself the menial role of a clinical clerk and mapped out a regimen of rigorous observations of disease and precise note-taking. After several years, his work gained the attention of the medical public in 1825, when he issued a large study of tuberculosis prepared according to what he called his "numerical method" of medical analysis.

Louis's method consisted of a series of elements: careful observation and description of details, systematic record-keep-

ing, rigorous analysis of multiple cases, cautious generalizations based solely upon observed facts, verification through
autopsies, and therapy based more on the curative power of
nature than on the deliberate practice of the art of medicine.
Using this method, Louis published other major works in the
next few years, particularly his study of typhoid fever, and
became influential in the Parisian medical galaxy. His prominence was ensured by his convincing refutation of the dogmatic Broussais and of Broussais's physiological system of
therapy. At the same time, Louis was subject to some ridicule
for his apparent downgrading of the image of physicians; his
making record-keeping a task for physicians seemed to many
doctors a radical and disagreeable step down in their traditional work as theorists and philosophers. However, in the
nineteenth-century hospital, such clerical functions took on
more and more importance for those devoted to the ideal of
bringing certainty into medicine.

Louis began to attract American medical students about
1830 precisely because of his reputation for "exactness of observation and talent for analysis and arrangement."[15] These
young men could have gone to Great Britain to study medicine, as their fathers had done, but as Henry I. Bowditch
noted, word quickly spread that, in contrast to an earlier day,
extended study in Britain now, in the 1830s, was time "worse
than wasted." There, unlike in Paris, foreign students had little
access to hospitals, and relatively few professors commanded
much respect.[16]

The roster of Louis's American students of the 1830s and
1840s has since become a familiar one, because it includes
many who were ultimately numbered among this country's
most noted and influential physicians. They came from every
part of the country. From Charleston, there were such men
as F. Peyre Porcher and Peter C. Gaillard; from the deep
South, John Y. Basset and Josiah Nott; from Philadelphia,
William W. Gerhard, Caspar W. Pennock, and Alfred Stillé;
from New York, Alonzo Clark and Valentine Mott; and from
Boston, Bowditch, James Jackson, Jr., Oliver Wendell Holmes,
and George C. Shattuck, to mention only a few.

The earliest wave of Americans enjoyed unusually close

relationships with Louis. Some of that band, like Gerhard and Jackson, arranged with the master for private instruction in diagnosis. Several joined Louis's Société Médicale d'Observation. A few stayed in the hospital wards with their teacher while the cholera epidemic of 1832 swept the city of Paris. Louis from the start gave extra privileges to these students who had traveled so far to learn from him. "Écoutez," he said, when Bowditch asked for permission to visit the wards in the afternoons, "you are a stranger and are to reside here only a short time, and therefore I will grant you what I could not to a Frenchman."[17] With the personal attentions that he extended outside as well as within the hospital, Louis became a genuine father-figure for the students, and he helped make their stay in Paris comfortable as well as professionally profitable.[18]

Although such special consideration inspired great devotion in the Americans, it was Louis's extreme dedication to science and to hard work that influenced them most. As a group, the American students were well conditioned to the idea of having to work hard. They had been brought up for the most part in families of achievers, who were themselves laboring enthusiastically for the perfection of the country's social, political, and commercial arrangements or for the achievement of the religious millennium. The students sublimated their considerable zeal in an equally dedicated pursuit of medical improvement. Reports from Paris indicated that in general the Americans who studied medicine there during the 1830s and 1840s demonstrated both "moral purity" and "incessant industry."[19] And there is reason to suppose that the reports were substantially correct; Louis would hardly have suffered either fools or sluggards around him. In fact, he was greatly impressed with the Americans' energy and unusual sense of purpose and went out of his way to single them out for praise.[20]

The hard work with Louis did more than train the Americans in a systematic method of investigation. It opened up entirely new medical vistas for them. Typically, Bowditch found that study under Louis led to a nobler and more demanding view of medicine than any he had earlier conceived of, one that required "the highest and most laborious exertions of which man is capable." It instilled a new "power" in the medical man, he felt, one that people back home in Boston could not imagine. Such training "clears up the difficulties which previously seemed insurmountable, lays open to your view what was before all chaos, gives order where there was no order before."[21]

Bowditch did not realize at the time, though he did years later, that there were some blind spots in the training provided by Louis. The intense preoccupation with pathology and diagnosis left little room for speculation about the causes of disease. The severe skepticism toward current therapeutics left little room for appreciating the expanding pharmacology of the day. The rigorous regimen of personal observation at the bedside and in the dissecting room left little time to know about, let alone master, other new tools of medical science, notably chemical analysis, experimental physiology, and microscopic observation.

Not the least of Louis's shortcomings, all of which had their effect in American medicine, were his limitations as a statistician. The numerical method was, in fact, a circumscribed approach to medical numbers. While he spoke vaguely of the virtue of large numbers of cases, Louis actually reached his conclusions from relatively small numbers. Relying on amassing exhaustive detail about selected individual cases, he had little if any concept of the significance of a numerical mean drawn from all relevant cases. He was, moreover, virtually oblivious to the exciting statistical developments in the Paris of his time. Bowditch, whose father had translated the great Laplace's *Mécanique Céleste,* had every entrée to the circle of Parisian mathematicians, but Louis saw no need for his protégé to learn probability theory. Some of the Americans did make themselves familiar with the public health statistics of Louis René Villermé and Alexandre Parent-Duchatelet, but Louis did not encourage them to go very far toward the comprehensive statistical view of man and society that was being advanced by the Belgian polymath L. A. Quetelet. Most seriously, though Jules Gavarret demonstrated in 1840 how to apply sophisticated mathematical concepts to medicine, Louis never seemed to learn from that work; at least he made no effort to turn his American students in the directions it suggested.[22]

Such inadequacies ultimately led the Parisian clinicians and their followers into a dead end.[23] But Louis and his generation can hardly be faulted for their omissions when their contributions were so substantial. For, in American medicine alone, the positive consequences of these contributions were enormous and far-reaching. What could possibly have been more valuable for antebellum medicine than the lessons which Louis's students absorbed in Paris: the critical view of ancient therapies and hypotheses, the increased reliance on nature in place of fallible art, the spirit of disciplined inquiry, the devotion to an objective methodology—in short, a new vision of medicine as a science?

Numerical Method Becomes Gospel

Louis's students returned to the United States as members of a medical elite whose knowledge of medicine was far superior to that of most American physicians. The authority lent by this training, and not least the returnees' knowledge of numerical method, thrust them immediately into the front ranks of America's struggle for medical certainty and excellence. As a group they thus worked hard at implanting their master's ideas and numerical method in the mainstream of American medicine. Several reviewed Louis's books, translated them, and issued American editions. Some, like Bowditch, vigorously defended Louis against hostile critics. Some, like Holmes, dedicated their first scientific tracts to Louis. All brought to American medicine a deep commitment to the healing powers of nature, a conviction that harmonized closely with a prominent theme of their generation's leading philosophers, artists, and poets—the nature celebration of Emerson, Thoreau, Thomas Cole, Bryant, and their peers.

Some of Louis's students became teachers in American medical schools and hospitals, where they made certain that the new generation of students at least knew what the numerical method was. Holmes, for one, told the members of his classes that they should be thankful to be "born into the period of medical statistics," and he urged them to conduct their medical practices just as successful merchants did, "by an exact arithmetical process."[24]

William W. Gerhard made a deliberate effort to instill "correct habits of analysis" in medical students. At Philadelphia's Blockley Hospital he went to great lengths to duplicate Louis's numerical methods in his own lectures on clinical medicine and pathology, and he designated students "to preserve a record of such facts as may be useful materials for the science of pathology." Recognizing the difficulties to the student in trying to absorb large amounts of medical data, he decided "to arrange the different forms of disease in accordance with the facts we now possess." Gerhard divided the diseases he discussed into two classes, the organic and the functional, and then subdivided them, first into the acute and the chronic, and then into the general and the local. He hastened to warn that he was not constructing any definitive nosology, merely a temporary arrangement to serve as a working tool. In the present state of knowledge, he cautioned, "nothing can be more arbitrary than the classification and even the nomenclature of many affections." However, it was obvious that, as pathological anatomy and differential diagnosis improved, the ability to classify diseases would improve correspondingly.[25]

The earliest American advocates of numerical medicine were steadily reinforced by converts from among their own medical students, and sometimes by their preceptors also, as well as by additional physicians back from studies in Paris. The Louis-inspired ideas were widely endorsed also by editors of medical journals, through whom numerical concepts percolated steadily to the rest of the literate and curious among the orthodox practitioners. As a result, when the American Medical Association was organized in the mid-1840s, its committees almost immediately began bombarding the membership with Baconian inquiries, and the physicians lost no time in filling the annual transactions with quantitative findings.

The heightened emphasis upon observation and experience studied systematically by numerical methods discredited and began to supersede most of the vestiges of the great theoretical systems, and even ardent medical systematists began to give at least lip service to quantitative methods. Medical observers who in an earlier generation would have turned to the authority of a Hoffmann, Cullen, Rush, or Broussais, now looked for authority in assembled collections of data. Doctors all over the country were asking themselves in the 1840s, "What other method can do as much?" In the South, physicians such as John LeConte of Savannah and Samuel Cartwright of Natchez urged "statistical medicine" upon their peers, while the Georgia practitioner E. M. Pendleton was certain that the chief clues to the "great principles of Medicine" in his time were coming from statistics. Northern supporters of numerical medicine were equally enthusiastic and even more numerous than those in the South. Speaking for them, Theophilus Mack of Buffalo agreed that in statistics lay the "only true mode of advancing science."[26]

A coterie of Boston's senior physicians were particularly influential in spreading the numerical philosophy, even though they had not studied under Louis. Among these men were such eminent individuals as James Jackson, Sr., John C. Warren, and Jacob Bigelow. Jackson, Sr., who was introduced to Louis's method by his son, considered the Frenchman's teachings as the fulfillment of his own thirty-year search for a medical investigator who actually practiced Baconian ideas of measuring, weighing, and numbering. Only Louis had taken the gigantic step necessary actually "to pursue the method of Bacon thoroughly and truly in the study of medicine."[27]

The conversion of Warren and Bigelow in the 1830s was undoubtedly helped along by the influence of the Jacksons, but they had been confirmed Baconians for some time. In 1830 Warren summarized his understanding of the rigorous character of this mode of scientific inquiry. It involved, he wrote, "first, the careful observation of facts—and secondly, the comparison of these facts so as to deduce from them the laws of the science in question," both steps to be aided or supplemented by processes of classifying and otherwise organizing the masses of facts.[28]

The fullest American exposition of the numerical point of view in medicine was formulated by Elisha Bartlett in 1844, shortly after his service as mayor of Lowell. In this work, Bartlett did not simply restate the well-known concepts of Louis; he incorporated also the far more sophisticated numerical vision of Louis's Parisian contemporary, Jules Gavarret. Gavarret, several generations ahead of his time, had outlined in 1840 some of the potentialities for mathematics in medical-statistical analysis, including the use of the calculus of probabilities.[29] Bartlett agreed wholeheartedly that such expanded uses of numbers would give doctors a new confidence and status as real scientists in working with their findings. They would help complete the eradication of a priori medical systems, would sort out clues to the etiology of disease, and would effectively test the efficacy of therapies. Bartlett concluded that observation, statistics, and mathematics were as requisite to medical discovery as to any other science. It was therefore incumbent upon physicians to make their medical inquiries as exact as those of the chemist or the astronomer. "It is only by the aid of these principles, legitimately applied, that most of the laws of our science are susceptible of being rigorously determined."[30]

Bartlett's book, *An Essay on the Philosophy of Medical Science,* rallied the most literate supporters of numerical method. Readers recognized that no American had ever proposed so comprehensive or refined a methodology of medical investigation. They saw the work as a constructive alternative both to outmoded orthodox theories and to the troubling unorthodox medical movements of the day. Josiah Nott praised it as "the most remarkable medical book yet written in this country," though he rightly observed that its advanced calculus gave it little or no chance of being adopted by antebellum physicians.[31]

Bartlett's work clearly represented an unrealistic ideal for its time. Recognizing this, many of its enthusiasts during the late 1840s scaled down some of their expectations for numerical method.[32] Nevertheless, largely through their endorsement of the method, America's "regular" physicians had already reached a new consensus of orthodox medical ideas, a set of beliefs widely known as "rational medicine."

4

Numerical Method and Rational Medicine

Rational medicine—also known as "rational therapeutics" or "conservative medicine"—was the logical culmination of a half-century of empiricism and skepticism about therapeutics, from Pinel and Laennec to Louis and John Forbes. In the United States the leading spokesmen included such well-known figures as Jacob Bigelow, Worthington Hooker, Augustus A. Gould, Benjamin E. Cotting, and Austin Flint.[1] By the 1850s these leaders clearly believed that a solid majority of informed orthodox practitioners had come over to the "rational" view, that drugs, bleeding, purging, and other active forms of therapy should be administered much more sparingly and cautiously than they had been before. Greater emphasis was now placed upon proper nutrition, careful nursing, and various forms of individual hygiene. The art in medical practice had become firmly subordinated to what was called "expectant therapy," to the healing power of nature. As Worthington Hooker summarized it, Rush's philosophy of active therapeutics—applying measures on an extraordinary, "heroic," scale in the hope of doing "positive good" to the sick person—had been supplanted, among "quite a large proportion of physicians, by the opposite desire, by Chomel's golden rule," to guard against doing the patient harm.[2] Meanwhile, as the scientific foundation of rational medicine, numerical methods of inquiry were expected to lead to new high levels of certainty in medical knowledge.

Research and Differential Diagnosis

Numerical studies had not yet in the 1830s led the individual practitioner of rational medicine to any specific new forms of therapy, but almost everyone was confident that their intensive, continuing application would eventually shed much light on medical problems. Hooker expressed the hope that before long the quest for certainty in medicine would become a vast democratic project involving the observations of a "multitude of practitioners, with full records of their individual experience." Only through such a widespread effort could the medical profession sift through the claims of the various therapies. The great aim of rational medicine, he thought, was "to discover the exact circumstances which admit of positive medication, and those which forbid it." No place was left for generalized and subjective impressions or for mere listings of extraordinary or successful cases. Rather, an exhaustive collection and disinterested analysis of all related facts was required.[3]

Such studies, Hooker granted, were difficult and time-consuming. But some investigators, such as Alfred Stillé, whose own work was recognized to be "in the full spirit of this inductive or Baconian philosophy," reveled in, rather than minimized, the labors of numerical analysis. Clearly, the Protestant work ethic had a place in the pursuit of medical improvement; Stillé pointedly wrote that doctors should regard their profession "as a field of toil, and not as a garden of amusement." To the generation of the 1840s and 1850s, no one seemed to be more industrious at medical research than the statistical analyst using either his own observed data or those of others. Thus Stillé himself was respected as a "conscientious scholar, who patiently collects the records of others' labours, carefully sifts the wheat from the chaff, and with judicial impartiality selects what is of greatest value from their opinions and their facts."[4]

At this time, facility in reporting observations numerically was felt to be a necessity for almost any orthodox physician who wanted to have scientific impact upon his profession. Difficult as some made numerical analysis out to be, it was nevertheless the most available and easily applied tool for clinical research, certainly simpler, less expensive, and more convenient than the laboratory. And, for that matter, it could be used in instances where laboratory analysis did not yet apply, such as in studies of revaccination, the pulse, or respiration.[5]

Louis had hoped that at least some of his protégés would be able to devote themselves to full-time clinical research using numerical method. He particularly urged this upon James Jackson, Jr., but Jackson died soon after returning to the United States.[6] In any case, American prejudices were such that the idea of full-time research was still entirely unthinkable. As Jackson, Sr., put it: "In this country [such a] course would have been so singular, as in a measure to separate him from other men. We are a business doing people. We are new. We have, as it were, but just landed on these uncultivated shores; there is a vast deal to be done; and he who will not be doing, must be set down as a drone."[7]

Apparently none of the other returning students of Louis thought of doing full-time research here. None even took on the role of the physician as a clerical observer to the extent that Louis had, although several of the American disciples and converts undertook a variety of important diagnostic studies using the master's method. A central medical challenge of the 1830s was the mare's nest of common diseases that were all too often lumped indiscriminately under the name *fevers*. Notwithstanding Rush's assertion that all fevers were manifestations of but a single disease, sorting, differentiating, and analyzing them seemed essential before any thought could be given to their rational treatment. Physicians trained in Louis's methods of investigation had particular success in clarifying the relationship between typhoid and typhus fevers.

At the Philadelphia Dispensary and the Philadelphia Hospital in the mid-1830s William Gerhard and Caspar Pennock devoted much time to examining the various so-called remittent and intermittent fevers of the Middle Atlantic States. Following suggestions made by Louis, they also studied the typhoid fever of Philadelphia and determined that it was the same disease Louis had observed in Paris. In 1836 a highly malignant fever, which proved to be typhus, broke out in the city and soon became epidemic. Displaying some characteristics of bronchitis and some of typhoid, it was poorly understood by local doctors. Gerhard and Pennock had an excellent opportunity to observe cases of the disease and to perform autopsies at the Philadelphia Hospital. When the epidemic was over they pooled their observations of 214 cases and analyzed them using Louis's methods. Their results, published under Gerhard's name, provided the earliest authoritative differentiation of typhus from typhoid.[8]

During the late 1830s numerical studies of typhoid fever were also undertaken by James Jackson, Sr., Enoch Hale, George C. Shattuck, and other New Englanders. In 1838 Jackson, Sr., analyzed as typhoid 303 cases that had been treated between 1821 and 1835 at the Massachusetts General Hospital under the name "continued fever." In 1839 Hale compared that hospital's 197 typhoid cases admitted between 1833 and 1839 with similar diseases reported by physicians in France, Switzerland, and Germany, but he found no adequate clinical data for Great Britain. "I know of no English physician," he commented, "who has investigated the fevers of his country with anything like the patient diligence of the French pathologists." Similarly, when Elisha Bartlett set about to prepare a systematic compendium of current knowledge about typhoid, typhus, and other fevers, he had the observations of Louis, Chomel, and Gabriel Andral to draw upon, together with those of the Americans Gerhard, Jackson, Hale, and Nathan Smith, but he found the British treatises on the subject "obsolete."[9]

Another common American disease that received quantitative clinical attention, though not yet as extensively as typhoid, was tuberculosis. The earliest study of "consumption" that was done in this country "after the manner of Laennec and Louis," was Samuel Morton's study during the early 1830s at the Philadelphia Almshouse dispensary. In 1835 Morton summed up what he had done with a sample of cases, using careful description, detailed recording of symptoms, critical analysis of treatment, and postmortem examinations. He saw his work, however, as only a small beginning; tuberculosis required many more numerical studies. In particular, he urged that topographical, meteorological, and other statistical studies be undertaken to complement the purely clinical investigations he had made.[10] Morton was surprised that in 1835 there was no American edition of Louis's work on tuberculosis. Bowditch translated it in 1836 as his contribution to advancing "the holy cause" of medical certainty through knowledge of the numerical method.[11]

One of the most articulate spokesmen of rational medicine about 1850 was Austin Flint, a brilliant and creative physician who had already earned a reputation as "the American Louis," or "the American Laennec." Flint conceived of numerical analysis as "a means of developing truth, but it is still more important as a means of confirming or disproving the results of ratiocination."[12] Acting upon this conviction, he outdistanced any American of his generation in carrying out quantitative clinical investigations.

Flint had been introduced to the numerical method during his training at the Harvard Medical School under the elder Jackson, Jacob Bigelow, and John C. Warren, but he made his clinical reputation not in Boston but Buffalo. There Flint collaborated between 1840 and 1860 with Charles A. Lee, Charles B. Coventry, Frank Hastings Hamilton, Sanford Hunt, and other physicians in launching a medical journal, a medical school, a local medical society, and in obtaining facilities for public medical care. These facilities, especially the wards of the Erie County Almshouse and of the Charity Hospital, gave physicians ample opportunities for clinical investigation.

Flint used these resources for studies which incorporated increasingly large numbers of observations. Starting with rubeola in 1840, he went on to examine quantitatively cases of dysentery, continued fevers, pneumonia, and other diseases.[13] His keen attention to differential diagnosis through observation led to his remarkable investigation in 1843 of a typhoid outbreak in the hamlet of North Boston, New York. His conclusion that some form of contagion was involved in the spread of the disease just barely failed to implicate contaminated water as the transmitter. A more definitive determination that water was indeed involved depended, he concluded, upon the further careful collection of facts.[14]

Quantitative Evaluation of Therapy

However important such diagnostic studies were, Flint, Gerhard, and other rational physicians were even more interested in knowing what their research might reveal about the value of various current therapies. Although few actual attempts were made at this time to evaluate given therapies numerically, the Paris-influenced elite nevertheless made such assessment the cornerstone of their efforts to elevate and reform the orthodox components of organized medicine. One of the most influential and comprehensive statements of such a program came from Jacob Bigelow.

Though not trained in Paris, Bigelow had, since the late 1820s, nonetheless regarded orthodox therapeutic practices with increasing skepticism. As Professor of Materia Medica at Harvard, he was particularly struck by the alarmingly "frequent failures" of drugs and other therapies, as evidenced in the mortality figures kept by hospitals and individual physicians. Such failures provided powerful arguments for a greater reliance on the healing power of nature. In any case, Bigelow thought that simply as a matter of self-preservation, organized medicine had to discover what value these remedies had. In an 1835 address, he advocated a vast effort to determine to what extent diseases were "really self-limited and how far they are controllable by any treatment." He called for an objective "experimental comparison," not only of individual drugs but of the various approaches to treatment—the expectant, the active, the physiological, and others. He hoped that individual physicians would examine the results of their private practices. But it was obvious that "hospitals and other public charities afford the most appropriate field for instituting such observations upon a large scale." Clearly the methodology of Louis made this feasible: "The aggregate of results, successful and unsuccessful . . . will give us a near approximate truth. . . . The *numerical* method employed by Louis . . . affords the means of as near an approach to certainty on this head, as the subject itself admits."[15]

James Jackson similarly noted that the physicians of his generation had tried everything in an attempt to cure typhoid but still had no clear idea of which measure really worked. Now Louis had demonstrated that the course of that disease was not affected by any therapy then in use. Jackson concluded that Louis's form of analysis certainly should be extended: "The numerical method is easily followed in stating the numbers who recover and the numbers who die of any disease, under various modes of treatment, and likewise the duration of the disease in those who recover."[16]

Of prime interest was the question of the benefit afforded by bloodletting. In 1835 Louis published his *Recherches sur les effets de la saignée dans quelques maladies inflammatoires*, which analyzed the effects of bloodletting at different stages of pleuropneumonia, erysipelas of the face, and angina tonsillaris, and raised questions about its indiscriminate use. Jackson promptly had the work translated and published, along with an appendix which included observations of his own use of bleeding in thirty-four cases of pleuropneumonia in Boston. His experience generally supported Louis's findings for that disease. Louis's observations, Jackson concluded, indicated that the "benefits derived from bleeding in the diseases he studied are not so great and striking as they have been represented by many teachers. If the same results should be obtained by others, after making observations as rigorous as those of M. Louis, many of us will be compelled to modify our former opinions." Not a blanket condemnation of bloodletting, Jackson's work, like Louis's, urged selective use of the measure and a continuing inquiry into its merits for treating other diseases. "Ten hospitals, under the care of honest physicians," Jackson thought, "may settle the questions discussed within five years." The editor of the *Boston Medical and Surgical Journal* concurred in 1836 and hoped that physicians everywhere would start keeping the necessary records of their uses of this popular therapy.[17] Ten years later, however, well-educated physicians still looked in vain for a firm consensus on the effectiveness of bloodletting. "Where, I ask, are the data?," lamented Flint. Where were the detailed comparisons of patients treated with and without a resort to bloodletting?[18]

One of the few significant antebellum studies of this therapy was by Pliny Earle. In the course of a continuing study of

American insane asylums, Earle found that in the twenty-five years since the death of Benjamin Rush, both British and American physicians had largely given up the once-standard and largely indiscriminate use of bloodletting in the asylums. Experience seemed to indicate that "the lancet has probably confirmed and perpetuated more cases of insanity than it has cured." Earle found confirmation of this during the 1840s and 1850s in the reports of virtually every American asylum. From his own cases at New York's Bloomingdale Asylum alone, he reported that the decreased use of bleeding had been accompanied by an increase in the number of cures.[19]

Somewhat more numerous than assessments of bloodletting were studies of drug therapy. Some of these involved simple numerical assessments of the action of given drugs on particular diseases. D. J. Cain of Charleston, concerned in the late fifties with the difficulties of ensuring adequate supplies of quinine, found in experiments on fifty-four patients, that chinoidine, another derivative of the "Peruvian bark" (cinchona), seemed to act "as promptly and as efficiently as Quinine" in the treatment of intermittent fever. In Philadelphia, J. C. Morris experimented in 1856 with forty-three orphan children at the Preston Retreat to determine the possible effect of belladonna in protecting against scarlatina; though he did not pretend to understand why, his figures at the end of an epidemic that winter showed that 75 percent of the unprotected children had come down with that fever but only 53 percent of those who had taken belladonna contracted the illness. In a different kind of analysis, James B. Colegrove's data seemed to show that alcohol was an important cause of edema which occurred with intermittent fever; out of fifty-eight cases of edema discovered at the Buffalo Almshouse, forty-two were in heavy drinkers.[20]

The proper dosages of various drugs were also investigated in some numerical studies. Despite the general trend toward less-frequent use of drugs, many doctors still tended, as Stephen Williams pointed out, to use large or even heroic doses when confronted with particularly stubborn diseases. Thus, in the years 1849–1853 the desperate attempts to control cholera included resorting to large doses of such remedies as laudanum, quinine, ether, or even conserve of roses, along with the old favorite, calomel. To be sure, many doctors favored only small doses of these substances, while some doubted that any drug helped very much.

Jacob Bigelow felt that there was "no sufficient reason for believing that calomel cures cholera in any doses, large or small." No one could yet muster the statistics to prove this, but Bigelow thought he knew why. "It is the bane of medical science," Bigelow concluded, "that physicians do not publish their unsuccessful cases—that scores of failures are suffered to pass without notice."[21]

Bigelow's contemporaries also thought that certain diseases other than cholera were attacked best by large doses of one remedy or another. George B. Wood, noticing a midcentury decline in deaths from tuberculosis listed in the Philadelphia bills of mortality, thought that the "liberal" prescription of cod liver oil, which had recently come into use, would lower the mortality rates from that disease even more. In the South and West, physicians treated typhus and other congestive fevers with either copious bleeding or large amounts of a remedy like tartar emetic. A. G. Henry claimed he had the statistics to prove the value of large doses of opium. Calomel, however, remained the great standby for the rural practitioners of those regions, even though enlightened physicians had come to recognize it as "rather a specific in destroying, than in saving life." Among them, William P. Hort of New Orleans criticized the old habit of invariably and indiscriminately prescribing calomel, and he estimated that its effect in the treatment of congestive fever was "to consign at least two thirds of the patients to the grave."[22]

Treatment of at least some diseases had shifted substantially by the 1830s to quinine sulfate, the new derivative of cinchona bark. Although some doctors were unsure of this powerful medication, many practitioners used it and were highly enthusiastic about its effectiveness. Apparently there were no extensive efforts in this country to evaluate the effects of quinine until applying it in massive doses was proposed. Subsequent investigations of that form of treatment were the only antebellum instances in which the numerical method was applied extensively and systematically to the assessment of a drug.

Thomas Fearn and W. J. Tuck were among several southern physicians who first began treating bilious and other fevers successfully with very large doses of quinine. By the late 1830s practitioners throughout the country were reporting numerically on their experience with the drug. In 1839 Austin Flint experimented with various quantities of quinine in treating forty people for intermittent fever. His data showed that large doses were not only safe but could be used without the usual emetics, cathartics, or other form of "preparatory" treatment. Flint thought that many more such demonstrations would be needed to overcome the prejudices that orthodox physicians had against administering large doses of this drug, but he noted that nostrum makers had no such inhibitions.[23]

About this same time, during the Seminole Indian wars, several army medical officers in Florida reported on successful uses of quinine in large doses. J. J. B. Wright administered the drug lavishly to about fifty soldiers who were suffering from "protean" malarial fevers, Charles McCormick and John B. Porter to several hundred each, and all with much success. In 1843, recognizing that the experience of a few individual physicians was far from conclusive, Surgeon General Thomas Lawson solicited further reports from all of his medical officers on their uses of quinine. Fifty-seven ultimately responded to this questionnaire. The replies, both qualitatively and quantitatively, overwhelmingly favored the continued use of large doses.[24] Bennet Dowler, who reviewed the reports, however, found many of the data imperfect and lacking in numerical precision; some of the replies, moreover, failed to take into account that deaths occurred among some of those being treated with the drug.[25]

Clinical Record-keeping

In the final analysis, the regulars had to admit what homeopaths already had been pointing out, that "of all departments of statistical medicine, treatment is the most uncertain." Mid-nineteenth-century American hopes of obtaining authoritative evaluations of quinine, calomel, bleeding, and other therapies had simply foundered at the very start upon the shoals of inadequate medical data. This failure was deeply frustrating to the new clinicians, who felt that adequate record-keeping was so central an adjunct both of hospitals and of private medical practices.[26]

By the 1840s there was little argument among orthodox medical spokesmen over the importance of good medical records. The ideal of medical truth, it was assumed, could never even be aspired to without meticulous recording of data: "Such records are the source of all the knowledge we possess. It is by them that we have become acquainted with the history of disease in different ages and countries; the appearance and the disappearance, the increase and the decrease of particular maladies, and the tendency of certain localities, professions, and modes of life to protect from or to expose to diseases of particular types; they are the source of all our knowledge of diagnosis and of prognosis, and they form the only convincing proofs of the efficacy of remedies."[27] In his pathology text, Alfred Stillé spelled out some of the ways of reducing observed clinical facts to written records as a step to their effective arrangement, analysis, and publication.[28] First, however, one had to obtain the data.

American hospitals in the 1840s were still frequently deficient in instituting the mechanisms necessary to ensure good records. Few could boast even the minimal procedures that Flint instituted at the Buffalo City Almshouse in about 1843 for recording clinical data. These included, he wrote, "a Register in which were entered the name of each person brought to the house sick, or taken sick in the house, the disease, habits, date of registration, and discharge or death, with occasional brief remarks. In addition to this, all the prescriptions were registered daily, and the details of some cases of peculiar interest recorded in a case-book. This duty was performed mostly by a resident pupil."[29]

Much more usual were the arrangements at Philadelphia's Blockley hospital, where in 1844, John P. Tabb could hardly find enough data to prepare his "inaugural thesis" for Jefferson Medical College. "Who would ever imagine," he commented, "that a death-book afforded the only records of a large hospital, yet such is the fact; ˙and it is only of late that any diagnosis of disease has been registered on the patient's admission."[30]

Thanks to the efforts of the new clinicians, this situation improved during the next few years. By the 1850s, in fact, the American hospitals seem to have gone beyond even those of Paris in keeping full day-by-day records of cases. Flint, during his stay in Paris in 1854, discovered only one hospital in which the routine, "careful registration of clinical observations" was being carried out. For Flint, as for other American physicians who had embraced Louis and his method so enthusiastically, it was a shock to realize that the clerical activities so important to the master were now being pursued less diligently in the hospitals of Paris than in the United States.[31]

Because most physicians in the 1840s and 1850s still performed the bulk of their work outside the hospital, the state of their private record-keeping habits was also a matter of concern. Since the desirability of extensive and careful records was a lesson that had to be hammered home repeatedly, editors, professors, medical society officers, and commercial publishers in all parts of the country kept busy stressing these needs. In areas near the frontier, there were special problems. Daniel Drake reported that disarray and disorder were the general rules in physicians' offices there, and even those who were interested in making any kind of medical inquiry tended to have habits of observation that were "extremely imperfect." Drake exhorted his students to at least serve as models for changing these deplorable habits when they started out in practice: "You will enroll [that is, associate] yourself with physicians and surgeons, who, on the main, record as negligently as they observe. I might rather say, that you will go forth among those who do not record at all; for the majority of the physicians of the Mississippi Valley keep no other books than those in which they write down the names and debts of their patients, now and then adding those of their diseases."[32]

By midcentury, publishers were advertising an increasing variety of office books for physicians. Some of these were ledgers to facilitate the collection of bills; others were visiting lists with handy tables of poisons, antidotes, and proportionate doses for different ages. Still others were designed to help the doctor systematically record every kind of medical detail about his patients and their complaints. The best of these record books provided from ten to twenty ruled columns in which to enter such diverse facts as race, occupation, and condition of the environment, as well as information on symptoms, diagnosis, medication, and outcome. If medical men had any inclination at all to make observations, here was the tool to enable them to "follow the example of Hippocrates in keeping a record of their cases."[33]

In midcentury the medical profession entertained few doubts that the accumulated data from thousands of such physicians' record books would eventually go far to illuminate the outstanding questions about therapeutics, etiology, and pathology. Already, in fact, the records of relatively small numbers of doctors were contributing to the analysis and improvement of some emerging medical specialties. Of these, none were more profoundly influenced by numerical method than obstetrics and surgery.

Obstetrics

Relatively little detailed study of this country's obstetric practices could have been expected before 1800, or as long as attendants at births normally were relatively unschooled midwives. By the mid-1830s, however, the midwife had been largely displaced by the male physician-accoucheur, at least among middle- and upper-class women in the northern states. In their attempts to improve their practices, many of these practitioners found some relatively simple form of quantitative analysis useful. Those whose studies were at all substantial found the numerical method peculiarly appropriate for obstetrics because "comparatively few circumstances are to be taken into the account."[34]

Medical societies began surveying the various obstetrical practices of their members, and individual physicians began publishing their experiences with deliveries in homes, in institutions, or in both.[35] In Alabama, W. H. Gantt tried to stimulate comparisons between the results of obstetrics practiced in the city with that practiced in rural areas. He was persuaded that sufficient data, gathered over a period of years, would refute the galling image city people held of "the country practitioner as little better than the old midwife."[36] Henry A. Ramsay's obstetrical records concerned mainly black women, since planters in his area of Georgia routinely called him for all deliveries of their slaves. When Ramsay published these figures in 1850, however, their accuracy was questioned before the Georgia Medical Society, and later they were suppressed in the American Medical Association.[37] Other practitioners around the country used their data to test the folk belief that the time of parturition was governed by changes of the moon, to demonstrate the effects on fecundity of particular climatic conditions, or to prove some other favorite hypothesis.[38]

During the decades before the Civil War obstetrical data moved steadily from mere collections of numbers of births, miscarriages, and abortions, to more complex information. Some physicians began to record such social data as the nationalities, ages, marital states, and numbers of previous pregnancies of the mothers. Others kept increasingly technical details: times of the day, month, or year of deliveries; the length of gestation or of labor; types of presentation; results of the use of forceps or other instruments; and weights of the children at birth.

Physicians used their statistics particularly for guidance in handling the various difficulties accompanying childbirth. Some doubtless agreed with the hydropathic physician R. T. Trall, whose figures showed that most accidents and fatalities in childbirth came about whenever midwives had been replaced by physicians, but the medical profession offered little support for such views and looked instead for ways to improve the outcomes of their deliveries. The hesitant introduction in 1850 of instruction in "demonstrative midwifery" using living subjects at the medical school in Buffalo was a major step forward.[39] Other improvements came from statistical analyses.

J. P. Maynard began to question the widespread treatment of puerperal convulsions by bloodletting when he noted that twenty percent of the mothers so treated died. When he found that seven cases in his own practice had recovered without bleeding, he called on the rest of the profession to collect a body of data that would definitely prove the wisdom of abandoning that treatment. In addition, several physicians tried to sort out numerically the various factors in the hemorrhaging of newborn infants and effective measures for controlling it. Other doctors were especially concerned by the high maternal mortality resulting from hemorrhage in placenta praevia. A committee of the American Medical Association in 1849 was certain that the old methods of handling this condition could be proved ineffective in reducing mortality, but it remained for James D. Trask to study the effects of these various methods as used in several hundred cases.[40]

Far more difficult to grapple with was the mortality from puerperal or childbed fever, an infection whose devastating past appearances in the wards of Paris's Maternité and other European institutions haunted every midnineteenth-century American proposal to build a lying-in hospital, and whose numbers of victims had damaged the reputation of many respected physicians in individual practice. Up to the 1830s the circumstances of outbreaks of the fever left ample room for honest differences of opinion as to its cause. Hugh Hodge searched the mortality records of the Pennsylvania Hospital for clues in 1833, but he could do no better than attribute the outbreaks there quite vaguely to changes in the condition of the atmosphere around the hospital. Charles D. Meigs, also of Philadelphia, concluded from his own practice, and from his reading on the subject, that the affliction was not even a fever but a nonspecific inflammation sometimes accompanying parturition.[41]

Though both views were influential, published reports of cases of puerperal fever during these decades increasingly suggested some form of contagion, specifically the physician himself, in transmitting the infection from one patient to another. In 1843 Oliver Wendell Holmes questioned Boston obstetricians on the matter and also searched published case reports for information. The evidence he gathered thoroughly convinced him that puerperal fever was contagious, so much so that he came to regard it as little short of criminal for the physician to neglect to take simple precautions against spreading the infection.[42]

A dozen years later, the continuing disbelief on the part of several leading obstetricians that contagion could play a role provoked Holmes into reissuing his paper on puerperal fever. Meigs, in particular, habitually ridiculed the idea of contagion in his classes at the University of Pennsylvania. Meigs was then attributing the spread of puerperal fever to chance, or to Providence. Equally provocative, perhaps, for someone like Holmes, was Meigs's offhand dismissal of the quantitative arguments in the case. Reviewers sometimes noticed that Meigs used tabular evidence when he needed support for some point or other, but his public position was that statistics "ought never to be the physician's guide. I am no admirer of medical statistics."[43]

Holmes, a professed admirer of them, strengthened the moral argument of his own article by stressing the authority of statistics. He called upon a statistical professional, an actuary, who readily ascertained that chance was "out of the question" as a reasonable explanation of the spread of puerperal fever, while there was at least "*some* relation of cause and effect between the physician's presence and the patient's disease." It was evident that the occurrence of as few as three cases of the disease closely following each other in a physician's practice was "*prima facie* evidence that he is the vehicle of contagion." Women about to be delivered by such doctors obviously were poor insurance risks. Moreover, Holmes repeated, physicians who ignored this evidence should be held liable for manslaughter. Holmes did not aspire to produce the same sort of original evidence that his Hungarian contemporary Ignaz Semmelweis had produced. He made no careful statistical studies of the incidence of puerperal fever in different Boston hospital populations, but by and large, Holmes's American medical readers seem to have been impressed and satisfied with the facts that he brought forward to support his indictment of contemporary obstetrical practice.[44]

Statistics and the Elevation of Surgery

If numerical methods made some contribution to the improved safety of mothers after childbirth, they proved to have an even greater practical impact upon the development of other surgical procedures. Before the Civil War, the extremely rapid pace of national growth was accompanied by an increased demand for surgery. In this climate of development American surgeons performed many more conventional operations than previously, and they were ingenious and daring in trying out new ones. With the examples of Ephraim McDowell's pioneer ovariotomies, William Beaumont's experiments on the stomach of Alexis St. Martin, and J. Marion Sims's repair of a vesico-vaginal fistula, to name only some operations, people were already boasting about the "glories" of American surgery. The expansion of the field, however, was viewed with apprehension by some surgeons who perceived with others that some measure of order or control over the growth was needed to ensure proper care for patients.

Quantitative analysis offered a promising means of evaluating the various surgical procedures that were being performed. It caught on quickly in the United States and proved to have more overall significance for the antebellum development of the specialty than any single surgical operation. Without much fanfare, the use of numbers became almost as natural to many surgeons as the use of the scalpel. By the early 1840s, analyses of accumulating figures began to provide guides to the utility, safety, and probable outcome of various operations, and thus to exert a healthy restraint on recklessness or guesswork in the use of some of the brilliant innovations of the day.

When Sir William Osler long ago drew the attention of medical historians to some of the American students of Louis, he did not stress those who had become surgeons. By this omission he inadvertently left the impression that antebellum surgeons were not interested in the numerical method. Nothing could have been further from the truth. The surgical amphitheaters of Paris were, of course, among the prime attractions for the foreign students who flocked to that city between 1820 and 1860. Americans interested in surgery studied under a talented body of operators—besides Larrey and Dupuytren, there were Philippe Roux, Armand Velpeau, and many more. However, most students spent time also with famous teachers in other fields, and certainly some, including several who achieved prominence in surgery before the Civil War, found their way to Louis's lectures.

One of these bright young men was George Norris of Philadelphia. While he was in Paris during the early 1830s Norris not only followed Louis's lectures avidly, but worked hard enough to win election to Louis's Society of Medical Observation. His subsequent American applications of the numerical method were so substantial that he gained special praise from the master. Various other students, men such as Jonathan Mason Warren of Boston and Willard Parker of New York, were equally influenced by Louis and went on to make extensive use of the numerical method in their surgical careers. In addition, experienced surgeons of the caliber of William Gibson and John Collins Warren, who traveled to Paris on vacation during these decades, took the trouble not only to visit the famous French surgeons but to seek out Louis in his wards at l'Hôtel Dieu. Warren wrote back that every medical student in the United States should learn Louis's numerical method before going on to any other phase of medical education.[45]

Through such individuals statistical methods were being introduced into the surgical side of American clinical medicine by the late 1830s and early 1840s. During the next two decades the taste for statistical analysis and reporting spread to surgeons in all parts of the United States. In hospitals, surgical staffs began to compile lists of the various kinds of surgical cases treated during a given year. In medical journals, articles on the statistics of surgery from private as well as hospital practice proliferated almost to the extent of articles on vital statistics. In medical societies, committees tried to outdo each other in circulating questionnaires aimed at gathering a wide variety of surgical data.

One of the first surgical applications of statistics was an attempt to assess the most feared operation of the day, amputation. Few other operations reflected so vividly the mounting hazards of the industrial revolution. Amputations recorded in a single surgeon's practice could well indicate the range of such calamities: hands "torn in a cotton carding machine," arms lacerated by gunshots, feet crushed in mining or excavating accidents, legs "mashed by a locomotive engine."[46]

Early in 1838, shortly after his return from Paris to Phila-delphia, George Norris made an initial compilation of the amputations performed at the Pennsylvania Hospital over the previous seven years. Norris wanted to demonstrate, first of all, that "contrary to the opinion generally prevalent in this country, amputation, even under favorable circumstances, is very frequently followed by fatal results." The record of twenty-one deaths among the fifty-five amputees amply bore this out. Norris also suggested that the hospital's conservative policy of postponing amputation an unusually long time might well have contributed to this high rate, but the data at hand were not adequate to shed much light on this matter. Another inter-esting question for Norris was just how surgery at the Penn-sylvania Hospital compared with that of hospitals in other cities and countries. He and his colleagues had the impression that, for one reason or another, the Pennsylvania surgeons had been able to save many badly injured limbs which would have been lost in most other hospitals, but few statistics for testing this belief were available in 1838, and none at all from other hospitals, either in this country or in Europe. In his surgical fact-finding efforts Norris was considerably ahead of his time.[47]

Norris's reliance upon numbers met with a certain amount of skepticism. This was particularly true of his conclusion that most writings of surgeons could not be trusted unless based on written records. Like their European colleagues, American surgeons traditionally relied on their memories rather than on records when they discussed their past practice. Honestly maintaining, from their best recollection, that they had never lost more than, say, one amputee out of twenty, they found it hard to believe Norris's statistics to the contrary. At least some agreed with the Georgia surgeon Paul Eve, moreover, that statistics of the kind Norris was gathering gave American sur-gery a bad name.[48]

Most medical editors, by contrast, quickly began to applaud Norris's data, along with his willingness, unusual for that day, to "bring out the unsuccessful cases into prominent relief, instead of keeping them carefully out of sight."[49] A few in-dividual physicians began to follow Norris's suggestion of keeping records of the amputations in their private practices.

Surgeons also began to publish the amputation statistics of other American hospitals—George Hayward for the Massachusetts General Hospital, Henry Buel for the New York Hospital. By 1848 Americans could compare the average mortality of 26.6 percent from all amputations performed in Philadelphia, New York, and Boston hospitals with the 26.2 percent in London and 53.4 percent in the nine Parisian hospitals studied by Joseph Malgaigne.[50] As it studied these death rates, the American medical profession gradually arrived at a somewhat better appreciation of the high risks their amputee patients faced regardless of the manual skill of the surgeon.

Statistics on fractures proved equally eye-opening. During the 1840s surgeons in several countries began to analyze the kinds of fractures being treated at particular hospitals. By 1851 Frederick Lente was thus able to compare the New York Hospital's figures with those of London's Middlesex Hospital and with Malgaigne's data for fractures treated at the Hôtel Dieu of Paris. Although the three institutions varied in their success rates for different kinds of fractures, none had grounds for self-congratulation.[51] Just why this was so became crystal clear with the findings, about this time, of the young surgeon Frank Hastings Hamilton, a colleague of Flint in Buffalo.

During the mid-1840s Hamilton took several months from his practice to travel in England and on the Continent. In Paris he spent much time visiting the various hospitals and attending lectures and operations of distinguished surgeons. Hamilton also went to see the famous Dupuytren museum, a rich collection of pathological specimens not far from the Ecole de Médecine. Hamilton took a professional surgeon's interest in the "diseases and broken bones" displayed there, and he spent many hours taking notes on the results of the many kinds of operations on fractures that had been performed by French surgeons. When he subsequently arranged his observations in tabular form, Hamilton was startled by what they revealed about the practice of surgery. The summaries showed that, out of a total of 180 fractures, only 55 had healed properly, leaving a shocking 125, or two-thirds, with some sort of deformity or irregularity. The results, he concluded, "must be regarded as testimonials of how little the very best surgical skill and appliances can accomplish under the most favorable circumstances, in the readjustment and retention of broken bones."[52]

Shaken by these findings, Hamilton spent much of the next decade in Buffalo systematically examining the outcomes of treatments of fractures. This meant the careful "measuring and 'leveling' " of every person he could find who had had a fracture in the past: checking his patients, those of other local physicians, and those who had been treated elsewhere, to determine where imperfect alignment, shortening, or other deformity had occurred. Between 1849 and 1855 Hamilton issued his findings in progressively more elaborate "fracture tables," which gave the average results of treatments of several hundred cases and amply confirmed his original finding that a very high proportion of fracture operations came out badly.[53]

Sobered by the evidence that their best efforts were meeting with such limited success, competent surgeons welcomed Hamilton's findings. The tables provided every surgeon with a means "to determine, pretty accurately, the chance for his patient," that is, the likely outcome of their current treatments. At less than ten cents a copy, Hamilton's pamphlet was recognized as a tremendous bargain for surgeons, particularly since it came out in a period when malpractice suits had become virtually epidemic. Many surgeons had laid themselves open to such suits by simply not being able to inform their patients ahead of time of the risks. Others became vulnerable to being sued using highly publicized instruments or techniques that were backed solely by their originators' extravagant claims. Only with Hamilton's figures could such claims be adequately verified. As one reviewer commented, "There is but one infallible test in all these cases—*let the limbs be measured, and let us have the statistics.*"[54]

Though the statistics on amputations and fractures were understandably conspicuous, hardly less attention was given to the numerical study of other surgical operations. Samuel D. Gross, concerned over the outcome of lithotomies, discovered that overall "about one out of every five" persons that were operated on for stones perished. At the same time, however, other Americans preferred to focus on the numbers of lithotomies performed and on the mortality records of the most successful American surgeons, a figure calculated in 1852 at only one death in every seventeen lithotomies.[55] Physicians were equally interested in the gross figures of ovariotomies performed, particularly because these operations had come to be considered preeminently "American." George H. Lyman dampened some of this provincial self-satisfaction during the mid-1850s by drawing attention to the low success rate of this operation—about 40 percent; his statistics clearly argued the need for more research before ovariotomies could become safe.[56] Similarly, the statistical studies of other surgeons—of John Shaw Billings on the surgical treatment of epilepsy; of Gross, Hamilton, and Jonathan Mason Warren on operations for malignancies; of George Lewis on appendectomies; of Bowditch on repairs of diaphragmatic hernia—all pointed to the continuing need for caution in almost every kind of surgery and frequently highlighted the need for more and better statistics.

One of the most important surgically related procedures requiring evaluation was anesthesiology. In fact, the medical profession turned to statistics for such assessment soon after the first operations utilizing ether were announced in 1846. Electrifying as this news was, many physicians initially viewed anesthesia warily, at least partly because of the "wild enthusiasm" of some of its advocates. Statistics on actual uses were certainly needed to prove its efficacy and safety.

Reports of such experiments were quick in coming. Jonathan Mason Warren early in 1847 published a study of the first 19 operations using ether at the Massachusetts General Hospital. In 1848 Walter Channing published his experience with ether in 581 cases of childbirth. In that same year other surgeons began to summarize their experience with chloroform, and some began to gather information on the comparative effects of the two anesthetics. By 1848 figures indicated that chloroform had already largely replaced ether, and the masses of favorable evidence for both had generally swept away the initial skepticism. Widespread trials, both in the United States and in Europe, showed that they posed little danger even when used under poorly controlled conditions. Although a need for continued testing of anesthesia remained, individual American surgeons, dentists, and obstetricians by and large had no more doubts about its safety at midcentury than did the thousands of citizens who had taken to sniffing chloroform recreationally.[57]

Statistical analysis helped raise antebellum American surgical practice to a level of respectability that compared favorably with that of European surgery. In fact, the accumulation of data showed American medical leaders at midcentury that extraordinary overall progress had taken place in every branch of their profession in the past fifty years. They could point to a steady increase in the numbers of hospitals, dispensaries, and clinics built in their lifetimes, to the rapid multiplication of native medical journals and texts, to the prolific spread of American medical schools. They could note the remarkable expansion of medical knowledge and the new ease of access to medical information. They could remind their audiences of the benefits brought by such innovations as vaccination, the stethoscope, quinine, and anesthesia.

Given such abundant progress and improvement in the medicine of their day, "regular" physicians might well have been expected to be enjoying substantial authority, security, and status in society. Unfortunately, this was not the case for very many. Indeed, to an increasing extent orthodox doctors were insecure professionals who devoted much of their energies and their statistics just to defending their form of medicine. To be sure, Louis's numerical method almost by itself provided the orthodox American medical establishment with a new, much-needed aura of scientific respectability. But the establishment needed much more than that, for "regular" or mainline medicine was under fire from the partisans of "irregular medicine," from a multitude of enthusiasts, each pushing forward some medical innovation or therapeutic option—mesmerism, phrenology, homeopathy, vegetable diet, acupuncture, or hydropathy, among others, but, most prominently during the thirties, botanical medicine.

During the Jacksonian era, whether orthodox medicine was worth saving became an open question. If it was, could it be

saved by elitist intellectuals working for reform, with quantitative methods, from within the establishment? For most people, matters of personal hygiene were of infinitely greater
concern than clinical medicine in hospitals. The statistics of
health and disease were matters of central concern to all. However, for large numbers of people the most relevant and convincing statistics of the day were those emanating not from
the numerically inclined orthodox physicians but from the
sources that regarded mainstream medicine as the enemy.

5

Alternative Medical Approaches

The emergence of an orthodox, scientific, and profession-ally self-aware medical establishment was only one side of America's concern for medicine and health in the early nine-teenth century. Accompanying it and assuming progressively greater significance, especially after 1825, was the parallel ev-olution of what may be termed a medical counterculture, made up of advocates of an assortment of hygienic beliefs, new ther-apies, and new medical systems. Some of these beliefs and innovations became institutionalized, but others remained the vaguely defined aspirations and practices of an amorphous and indeterminate body of individuals. Some of the concepts at one time or another were tentatively embraced by orthodox medicine, and a few were incorporated into it, but most were eventually repudiated and faded away. Nevertheless, during the period before the Civil War they were collectively vigorous. As creative elements in the medical scene, they effectively chal-lenged the status quo by providing Americans with a wide range of therapeutic choices.

Like the medical regulars, members of the counterculture were ambivalent about the use of statistics. For some, their faith in the effectiveness of a given therapeutic system was strong enough to make numerical observation and analysis seem unnecessary. Others, however, found statistics as im-portant as regular physicians did, and used them extensively and effectively.

Jacksonian Democracy and Medical Pluralism

The medical counterculture brought together many who were privately or publicly dissatisfied with the conventional medicine of the day. It included physicians frustrated by the crudeness and ineffectiveness of regular therapies, along with a variety of individuals whose passion for abolition, Sunday schools, penal reform, or some other social goal somehow spilled over into medical reform. It included many who, having been swept up by the ideal of educating the democracy, found that improving the people's knowledge of hygiene was as essential as other kinds of instruction. It incorporated some who were being influenced in one way or another by revolutionary concepts in health care, as well as in politics, economics, and social ideas from abroad. A number of the members of the counterculture were middle- or upper-class intellectuals who were swayed by the Romantic outlook which gave the individual as much responsibility for his own health as for his spiritual and intellectual destiny. A much larger proportion was composed of humbler folks—antimonopoly democrats, antirent agrarians, and others who resented accumulations of wealth, privilege, or authority, whether manifested in physicians or in lawyers, bankers, or clergymen.

The medical and social worlds of the Jacksonian common man were as far removed as could be from those of the foreign-educated clinicians. For those who were illiterate, ignorant, or gullible, neither science nor statistics meant very much. For those accustomed to follow old wives' tales and folk remedies, the extensive formal medical progress of the Enlightenment might as well not have happened. Individuals of the so-called lower classes—poor farmers, town laborers, slaves, and others—but also the credulous of the middle and upper classes—patronized the country's wandering healers, nostrum peddlers, and quacks, and not a few of them believed firmly in medical astrology, including numerology. Of course, when used as a guide to medical or other action, numerology involved making simple arithmetic calculations and keeping a few home records, but the mysterious qualities it ascribed to numbers were foreign to the basic factual characteristics of statistics.

Throughout the period, physicians noticed that astrological beliefs were frequently practiced in their communities, and repeatedly doctors felt it necessary to help discredit popular attribution of epidemics to such unusual phenomena as comets, earthquakes, or large migrations of birds. Some of them tried to test the doctrine that certain days were critical to medical well-being, and some produced statistical evidence that seemed to prove it. The University of Virginia physician Robley Dunglison thought that credence in solar and lunar influences on the body had declined greatly by the 1830s, but he admitted that there was a continuing belief that the moon was a cause of insanity. Dr. Stephen Williams compiled a table from sixty-six years of mortality records in Deerfield, Massachusetts, to prove that there was no special danger from the supposedly fatal "climacteric years," that is, years that were multiples of seven. Thomas Hersey thought it preposterous to attribute menstruation to astrological phases of the moon; if true, all women in a given area would menstruate at the same time, creating "a sort of universal flood-tide."[1]

One Jacksonian writer, however, convinced that men as well as women had monthly indispositions, proposed that dispensary physicians should gather statistical data on the phenomenon. In any case, the public ought to be warned of some of its potential social consequences. "Elections should never be held on these days of malaise. [Neither] Balls, christenings, nor any of the great affairs of life, should be conducted on days when one half of the women are out of sorts, and men in the dumps."[2]

Not everyone at the grass roots of America, of course, was superstitious or ignorant, and the basic assumptions of republican life challenged these traits by placing a premium on education. Within a framework of expanding opportunity and individual rights, the dominant tendency was to improve oneself and to avoid the onus of ignorance. In the context of medicine, this meant not only grasping whatever opportunities there were to learn about the body and its disorders, but utilizing whatever therapies were offered, whether within orthodox medicine or outside. The pluralistic countercultural medicine that emerged in the 1820s and 1830s was thus solidly based in democratic precepts. It stood for such things as broader health education and better medical attention for all citizens, for the abolition of medical privilege, and for the freedom to choose from among competing medical sects.

Benjamin Rush was claiming by 1809 that medicine in the new republic had already cast off many earlier ways which he regarded as having been "royalist." Medicines, he noted, were no longer withheld from patients whose cases were hopeless, and the sick were now informed both of their illnesses and of the nature of the drugs given them. Ostentatious dress and affected manners of physicians had begun to disappear, "and prescriptions are no longer delivered with the pomp and authority of edicts."[3]

A decade later Samuel Latham Mitchill concluded that broad economic opportunity and democratic ideals in action were actually creating highly favorable conditions for a pluralistic medicine. Nowhere on earth, he supposed, was there so widespread an appreciation for any kind of healer—root doctors, witch doctors, Indian doctors, and "obi," or African doctors, as well as conventional physicians. The access to medical care of one sort or another was one of the basic expectations of Americans.[4]

Not all physicians, to be sure, were enchanted by the progressive extension of democracy. One medical observer even suggested in 1820 that egalitarian ambitions had unfortunate medical consequences. Sometimes, he wrote, they led to "immoderate and ruinous expenses in style, furniture, and dress" among lower-class citizens, and the resulting impoverishment contributed to high death rates for their children. Nevertheless, by the Jacksonian period it was regarded as professional folly for physicians to criticize "lower" classes or to set themselves conspicuously above any segment of society. As one editor noted, if the average physician hoped to succeed "in a republican country," he must refrain from appearing condescending, haughty, or superior. Another observed that ultimately the doctor would have no choice but to submit his medical theories to the democratic processes: "Our country, by its free institutions brings every question, as it ought, before the tribunal of public opinion . . . Masonry has submitted to it—Mormonism, Agrarianism, Abolitionism, must all come to this at last, [while] *medicine*, like every useful science, should be thrown open to the observation and study of all."[5]

Popular Medicine and Thomsonianism

The botanical system of Samuel Thomson was an attempt to do just that, to make the ultimate equation of medicine with democracy. Thomson's system emerged as part of a long tradition of providing medical advice to laymen on how to treat themselves. The tradition included the preparation of tracts aiming to meet the medical needs of travelers, sailors, isolated families along the frontier, or anyone who could not afford to consult a physician. In places where physicians were in short supply, these works enabled large numbers of the population to be their own physicians. However, even when the supply of trained physicians began to meet the demand in the 1830s, the spreading egalitarian attitudes and growing suspicions about the value of conventional therapy helped ensure the continued flourishing of "do-it-yourself" medicine.

Samuel Thomson was a poorly educated farmer who learned the medicinal qualities of certain indigenous plants from "root doctors" in rural New England. In the course of treating his family and neighbors over a period of years he gradually shaped his knowledge into a crude therapeutic system, for which he managed in 1813 to obtain a patent. Having thus gained legal sanction in lieu of medical respectability, he proceeded to license individuals to practice his system at $20 per person. By the early 1820s practitioners who had purchased these rights were being enrolled in "friendly botanical societies," while they also received the mature version of Thomson's system in published form, entitled *A New Guide to Health; or, Botanic Family Physician.*[6]

The spread of the Thomsonian doctrine through this book obviously counted on large numbers of people being able to read. Within another ten years or so, these numbers were sufficiently great to lead to some institutionalization of the system. By about 1832 the movement had its own journal, *The Thomsonian Recorder*, a number of state and local societies, an annual national convention, and some 167 authorized agents who not only distributed rights to individuals and families but also sold medicines and ran infirmaries where the system was practiced. The movement reached its peak in the late 1830s, but then splintered badly because of differences between Thomson and some of his followers over such matters as profit sharing and the founding of botanical medical schools.

Thomsonian principles rested basically upon the founder's understanding (from the work of William Buchan and other popular medical writers) of the classical humors of the body in the context of a simple unitary view of disease. Somewhat like Rush, Thomson felt that, in theory, "all disease is the effect of one general cause and may be removed by one general remedy." But in practice he drew on a number of remedies; he was also shocked by Rush's extensive use of bloodletting and purgatives. The botanic armamentarium, by contrast, was originally composed of six vegetable preparations (most prominent of which were lobelia and cayenne pepper), used together according to specific formulas, which Thomson's handbook often identified simply by number. For the Thomsonians, heavy doses of lobelia and cayenne, along with a regimen of steaming the body to cause sweating, became as much a matter of faith as did bleeding and calomel for the followers of Rush. Dogmatic and self-righteous, in pursuing this faith, the Thomsonians became anathema to the regular profession. Almost alone among prominent regulars, the physician Benjamin Waterhouse endorsed the botanical regimen, thereby earning the eternal gratitude of Thomsonians and the eternal scorn of orthodox physicians.[7]

The Thomsonian movement's opposition to medical exclusiveness through its promotion of expanded personal medical practice among the masses elicited a sympathetic popular response. Just how widespread the system became is unclear. Thomson claimed in 1839 that the 100,000 people who had bought his book had made the system available to some three million more.[8] Independent observers, including some regular physicians, suggested that these figures may not have been far from the truth, particularly in view of the sect's extraordinary popularity in the Southern and Western states.

Whatever the true figure, it became apparent that Thomson's ideal of a universal do-it-yourself medical system was unattainable. As plain and simple as Thomson had made his system, "capable of being understood by every one," experience was showing that many people simply could not be their own physicians. Although the common person was still presumed to be intelligent enough, Thomsonians came to realize "that a great number will not take the necessary trouble." Faced with this basic problem, some Thomsonians thus comprehended by the late 1830s that to perpetuate their system they would have to have medical schools in which to train their practitioners adequately.[9]

Initially, like enthusiastic followers of many other causes, Thomsonians found little need for statistical ammunition to publicize or advance their movement. Rather, as believers in a newly proclaimed medical religion, they were carried along mainly on the strength of Thomson's own well-tuned appeal to rural and lower-class prejudices, together with the eventual accumulations of testimonials and sentimental verses that filled the various botanical journals. Though Thomson (unlike, for example, Joseph Smith and other contemporaries) never claimed a messianic role, he did not discourage people from regarding his work as "scripture." Scripture obviously needed no rational justification and could not be evaluated by usual medical or numerical standards. Waterhouse himself argued that "to weigh Patriarch Thomson in the scales of the regular physician, would be as unjust as for them to be weighed by his steelyards," or scales, and he saw each—Thomson and the regular—making his own contribution to medicine.[10] Nevertheless, Thomsonians lost no occasion to contrast the merits of their therapy with the faults of the "learned quackery" of regular physicians, and they sometimes tried to prove these merits with figures.

In 1832 and 1833 cholera spread far and wide across the country, and not a few Thomsonians welcomed the occasion to test the efficacy of the Botanic system. When orthodox physicians failed with their bleeding, calomel, and other means to effect cures, many of the sick turned to the Thomsonians, allegedly with great success. To be sure, in Columbus, Ohio, the board of health reported that by the later stages of the epidemic extraordinary mortality among families treated by the Thomsonians had driven most people back to the regulars. However, in New York City, botanics claimed to have saved 4,978 out of 5,000 people treated for cholera and asserted that the regulars had lost 1 out of every 2 patients. In Mississippi, where Thomsonianism was still illegal in 1833, newspapers reported that, out of 1,500 slaves on the plantation of Wade Hampton, over 700 had succumbed to the epidemic. What an argument for legalizing the new system, exulted the botanics. "Huzza for the Regular Doctors, Calomel, and the Lancet!"[11]

In the years following the epidemic, apparently few Thomsonian practitioners kept records of their practices. One who did, J. B. Woodward of Fall River, Massachusetts, analyzed the incidence of all disease in his city in late 1841 and early 1842. Out of 100 deaths, he admitted that 5 had been among his patients, but he attributed the remaining 95 to the drastic measures of the city's nine regular physicians. Individuals who conducted Thomsonian infirmaries seemed to be among the more active record-keepers in the sect, and at least one botanical editor used the rosy reports from those institutions to attack the mortality record of the city almshouses, in which mainline medicine was practiced.[12]

In general, however, most of those who spoke on behalf of botanic medicine argued not from concrete data so much as from generalized claims that thousands were annually saved by it. Conversely, the therapies of the regular profession were characterized simply as "powerful agents in the destruction of human life and happiness." Genuinely distressed by the conditions of neighbors treated by orthodox physicians, Thomsonians focused their hostility especially on the use of calomel, which they rightly believed inflamed the mouth and rotted the facial bones. To them, it seemed evident that "the records of death bear testimony" against the physicians who prescribed that drug. Thomsonians were completely serious in their insistence in 1839 that New Yorkers would need far fewer undertakers than their current number of 173 if they did not have some 497 regular physicians.[13]

Orthodox Responses to Quackery

The sporadic figures and impressionistic views published by the early botanic practitioners had little in common with the systematic data compiled by Louis's followers. The two groups were divided deeply by differences in methodology, background, status, and aspirations. To be sure, they shared a mutual opposition to indiscriminate bleeding and drugging, and occasionally a regular physician surfaced who had an open mind about Thomsonians and other sects. Gouverneur Emerson, for one, in 1827 urged other doctors to use statistics "as a means of estimating the comparative value" of the various systems of medicine.[14] By and large, however, regulars saw little need for such an exercise, since they had already made up their minds that botanical practitioners were quacks or empirics.

The struggle against quacks had been, of course, a constant factor in the efforts of all early nineteenth-century American physicians to organize, upgrade, and institutionalize their profession. Many states had adopted legislation for the licensing of medical practitioners, a clear indication of the regulars' success. The rise of Thomsonianism in the 1820s and 1830s, however, had helped bring about a whole new set of medical conditions. As the local botanical societies grew, they gained the voting strength, in state after state, to help force the repeal of the medical licensing laws and thereby win for Thomsonians and other irregulars the right to practice medicine.[15] Moreover, the new botanical institutions which began to appear in the late thirties entered into direct economic competition with regular institutions, and particularly with the orthodox medical schools. The rise of the Thomsonians also seemed to stimulate a great variety of other healers, both organized and unorganized, to become more active. Their scientific and economic base thus threatened, the antebellum medical establishment was forced to spend huge amounts of time and energy in the fight against irregulars and quacks.

Through these decades the regulars complained bitterly about the advertisements of assorted "quacks, jugglers, and conjurors" that filled the daily and weekly newspapers. As they watched the growing popularity of Swain's Panacea and other patent medicines, there was little they could do about the manufacturers' extravagant statistical claims that such concoctions were curing thousands of cases of disease. In Cincinnati, Daniel Drake reported that by the 1840s about as many warehouses were being required to house the huge supplies of "decoctions, balsams, elixers, lozenges, tinctures, honies, plasters, and linctuses" destined for drug retailers as were required by the city's large pork industry. In Boston, observers noted "tons upon tons of powders and boluses [being] sent off through the regular channels of trade," which then included, by actual count, eighty-nine apothecaries and forty-four druggists in that city alone.

Seemingly endless numbers of all kinds of medical entrepreneurs spread out over the land. Regular physicians were beleaguered not just by the Thomsonians but by swarms of homeopaths, phrenologists, and Grahamites; by purveyors of roots and herbs, Indian and Negro recipes, and patent medicines; by traveling dentists, bone setters, and mesmerists—all of them keeping step with the wandering peddlers, circuit riders, and newly fashionable makers of daguerrotype likenesses. It was bad enough that the quacks and empirics tried to give the regulars a bad name, but even more painful was the knowledge that the quacks would often "obtain better fees, and more prompt payment, than surgeons and physicians of acknowledged skill and elevated social position, and speedily accumulate property." It was a rare community that could claim to be "a place without quacks."[16]

Thomsonians and other anti-establishment groups offered the regulars only the most elusive of targets, at least until they began to organize institutions. Regular physicians could ridicule the botanics' use of Thomson's six numbered vegetable preparations and would accuse them of applying the concoctions mechanically. The Thomsonians, in turn, wasted relatively little time or effort in trying to explain the numerical aspects of their system and simply attacked the orthodox therapies outright. Since those therapies were increasingly vulnerable to criticism, by the late thirties the regulars despaired of finding any means to counter the Thomsonians and to regain some of their lost practice and lost public confidence.[17]

While the long-term answer to this situation lay in improving their remedies and professional organizations, mainstream physicians could take at least short-term comfort in some statistics from Southern cities detrimental to the irregulars.

In Natchez, Mississippi, then a town of some 6,000 people, the orthodox physician Samuel Cartwright produced data to show that a marked increase in mortality occurred following the 1833 repeal of the Mississippi medical licensing laws. Analyzing sextons' reports to the city council, Cartwright concluded that during the ten years up to 1833, when the town had only "regular" physicians, only 1 death occurred for every 61 residents, and only 1 in 30.6 even when boatmen and other itinerants were counted in. With repeal of the licensing laws a veritable "host of empirics," mostly Thomsonians, entered the town and began practicing medicine. The mortality immediately jumped to an annual average of 1 in 22.2 over the next five years. Cartwright did not, to be sure, claim that all of the increase was directly due to the scalding baths or lobelia administered by the empirics. He attributed most of it, in fact, to the "destruction of public confidence" in the old remedies of regular medicine. Nevertheless, he considered the data damaging to the so-called reformers: statistics, he concluded, "show that the syren voice of the empiric is not to be trusted."[18]

In Charleston, South Carolina, the "invasion of the Botanicals" was considered to have occurred in 1836, though repeal of the state licensing law did not come about until 1838. Much like Cartwright in Natchez, the physician Thomas Logan regarded Charleston before 1836 as an exceedingly healthy city, with a mortality averaging only 1 in 49 for the years 1832–1835. However, during the four years following the invasion, he found that the mortality soared to an average of 1 in 36. The new permissiveness toward quacks and empirics, he concluded, was a "disgrace to our country—a blot upon the enlightened age in which we live."[19]

Physicians in the North as well as the South welcomed the analyses of Cartwright and Logan, and many agreed that statistics provided the only effective refutation of the irregulars. One Northern doctor, however, Charles Knowlton, pointed out that the rough mortality figures of the two Southerners did not really provide conclusive evidence against the empirics. What was required were clinical experiments which, he facetiously observed to an audience in rural Shelburne Falls, Massachusetts, might go something like this: "Let a thousand cases, including all sorts of diseases, be put into a hospital, and there be treated entirely on the Thomsonian plan until all are either dead or cured; and at the same time let another thousand similar in all respects to the first be put into another hospital, and there treated on the regular plan until all are dead or cured, and we might then, perhaps, be in possession of statistical evidence that would be satisfactory to all."[20]

Since such a procedure was obviously impractical, physicians were left with the problem of empiricism still unresolved, and in fact, nothing the regulars could do at this time could really hide the apparent or real successes of botanical practitioners. In Columbus, Ohio, botanical spokesmen reported that 166 out of the 170 persons buried in the city in 1840 had been patients of the seven leading regular physicians, while only the remaining 4 had been under the care of the town's single Thomsonian "steamer." The loss of patients by the regulars was thus loudly proclaimed to have been 8 for every 1 lost by the steamer, though no figures were offered as to the overall clienteles of the various practitioners.[21]

One acknowledged success of the botanical practitioners was their campaign against calomel. By the 1840s many orthodox physicians were being forced to modify their prescriptions of that drug. In Alabama, many of the regular physicians who during the 1820s and 1830s had routinely "assumed that calomel was the *only* remedy for malignant autumnal fever" were reported to have lost many of their patients. Daniel Drake could find few "calomel and aloes doctors" left by 1840. Their declining success had the result, Drake thought, of bringing "into favor the botanical or steam doctors, whose methods were certainly preferable. Thus it is that eccentricity favors the growth of empiricism."[22]

Hygiene and Health Reform

The growing hostility of medical regulars toward Thomsonianism also extended to a considerable extent to those outside the establishment who were emphasizing age-old precepts of personal health. At least, when those precepts became the framework for alternative approaches to medical care, not a few regulars came to view hygiene as a form of competition and therefore something to be discouraged or rejected. To be sure, not all regulars were alienated from this side of medicine. Some worked hard to make individual hygiene or practical physiology a more prominent part of orthodox practice, to give the prevention of disease as much a role as curative medicine. Others were swept up in a broad and vigorous middle-class health reform movement that was hoped would virtually eliminate the need for curative medicine.

In the perfectionistic climate of the Jacksonian era, the good health of the individual citizen was widely held to be an attainable and worthy goal. If crime, irreligion, idleness, and other blemishes were unacceptable to society, equally so was the misuse of the human body and mind. Just as undesirable moral traits and social habits were to be identified and eliminated, so the unhygienic practices of the individual had to be measured and replaced by a methodical regimen that would ensure good health and a respectably long life span.

America's Founding Fathers and their physicians were familiar with the classical hygiene precepts of Cato, Bacon, and Luigi Cornaro, long before Sir John Sinclair reemphasized them in his *Code of Health and Longevity* in 1807.[23] Few up to that time, however, had as much appreciation as Sir John of the role that statistics could play in making clear the consequences of poor hygiene. Sinclair's collected precepts and statistics provided ammunition for a generation or more of lecturers on hygiene and physiology on both sides of the Atlantic. By the late 1820s, however, these educators were beginning to utilize newer knowledge about physiology and newer statistics on hygiene and longevity. Along with the numerical method of clinical medicine, a good deal of this new information came from Paris, but the Americans who contributed to health reform also produced more and more literature on these subjects, some of it backed by original data.

Health reform was, to some extent, a part of the introspective effort of the middle classes to achieve self-reliance.

Jacksonians prized self-reliance in health as well as in business, education, or religious faith, and "Know thyself" became as much of a practical motto for the generation of antebellum health reformers as it was a philosophical theme for Ralph Waldo Emerson and his fellow Transcendentalists. This self-scrutiny included much of the moral and spiritual introspection that generations of Puritans had become used to, and it also involved close observation of such personal matters as eating and drinking habits, styles of clothing, exercise, the emotions, sex, sleeping, and bathing.

As the middle class expanded and the industrial revolution got under way, more and more questions were raised about the effects of urban life on people. It was still argued that, with the advantages of their good climate, free institutions, and abundant food, Americans should be the most vigorous people in the world. But town life seemed to pose some serious obstacles to this ideal state. Among these was a rapidly expanding population of stay-at-homes. It seemed as if the whole generation of clergymen, professors, and other scholars now spent most of their time in their studies; as if more and more clerks, shopkeepers, and factory operatives were passing their lives indoors; and as if housewives rarely got out any more except to go to church and an occasional social event. None of these people appeared to be healthy. In the 1830s, the British traveler Harriet Martineau could find instances of really vigorous health only in the Alleghenies, in Michigan, and "among the ladies of Charleston, who pass three quarters of the year in the open air of their piazzas." Similarly, some of Emerson's friends in Concord thought that women of the period had generally "degenerated in strength," that American men were "a puny race," and that even "the cows are smaller."[24]

Such criticisms redoubled the determination of the Jacksonians to improve hygienic practices. They did not have anything very new to suggest, but they pursued the old principles more energetically and systematically than ever before. Quite apart from statistics on intemperance, the numerical accounts of the shortcomings in personal hygiene were weapons that almost every kind of social reformer could wield. Thus clergymen, teachers, physicians, and lyceum lecturers in every community used numerical facts to promote more healthful lives. Because of their efforts, larger cities began to build gymnasiums, personal sports such as walking and horseback riding were encouraged, and magazines sprang up to spread hygienics more widely.

Of the sympathetic regular physicians, John Bell and D. Francis Condie were among those who published popular health journals. Characteristically, they helped fill their columns with quantitative data on physiology and hygiene taken from the Parisian *Annales d'hygiène et de médecine légale*.[25] Bell, Edward Jarvis, John H. Griscom, and other regulars also prepared texts on hygiene, not only for their fellow doctors but for lay adults and for school children. The need for such works was demonstrated in 1842 by the educator Horace Mann, who found, among Massachusetts school students, that 10,177 were taking American history and nearly 4,000 algebra or bookkeeping, but only 416 were taking physiology.[26]

There were many laymen and medical mavericks who pursued hygiene as a holy cause largely outside the regular medical establishment. Concerning itself primarily with the health of the middle classes, this group of reformers also produced its own texts and journals, and, to reach as broad an audience as possible its lectures on physiology were often delivered by women as well as men. Thomsonians and homeopaths strongly endorsed this lay activity, partly because it represented a repudiation of mainstream therapeutics. Meanwhile, phrenologists, following the example of George Combe, made instruction in physiology and hygiene a central part of their system; Lorenzo and Orson Fowler subsequently made a fortune writing and publishing popular works on every aspect of these subjects. In the process, large numbers of Americans formed their basic hygienic habits without the direct aid of a regular physician.

The two individuals who set the tone and gave much of the early impetus to the popular health-reform movement were Sylvester Graham and William A. Alcott. Graham, a temperance preacher in New Jersey and Philadelphia during the early 1830s, gradually expanded his lectures to include a considerable range of information on health generally, including special advice to young men on sexual hygiene. By 1835 these ideas had evolved into the "Graham System of Living," a spartan regime involving abundant exercise, strict attention to clothing, cleanliness, and sleep, and dedication to a simple vegetarian diet, including emphasis on "Graham Flour." Adherence to such principles, he maintained, would ensure to the individual the ultimate quantity of personal well-being, "not only the longest life, but also the highest degree of health, and the greatest amount of happiness." There was, of course, a great deal to do before Jacksonian society would begin to approach such perfection. In fact, as Graham looked at the hygienic state of the United States, his words took on a good deal of the millennial spirit of the times. Like many individuals of the previous generation, he painted for his audiences a lurid picture of a degraded and declining race which had abandoned its original simple way of life, had turned shamelessly to a debilitating love of luxury, and had adopted such damaging dietary habits that its democratic ideals, indeed, its entire future, were threatened and undermined. American life expectancy now stood in stark contrast, he thought, to that of the earlier ages, when people had lived "an almost incredible length of years." In fact, the recent statistical evidence from American cities showed a decline in the traditional average life span of three score and ten to one of scarcely thirty years. "[This short life span] may be proved by any one who will examine our annual bills of mortality. Of all new-born infants, one out of four dies the first year. Three out of five only attain the fifth year, and, before the twenty-second year, one-half the generation is consigned to the grave. In our large cities, however, one-half die before they reach the tenth year."[27]

For the health reformer, the key to improving these dismaying figures was education of the people, and William Alcott was even more of an educator than Graham. Beginning in the mid-1830s Alcott, for a quarter century, played an intensely moralistic role as "missionary of health" to his generation, preparing an endless stream of books for laymen on health subjects and crisscrossing the country tirelessly as a public lecturer on the science of human life. As a physician, he gained popular confidence by insisting, even more strongly than Jarvis or Bell, that hygiene should become the central focus of medicine, that the ideal of prevention should replace the reliance on cure.

Henry David Thoreau noticed that many of his neighbors near Concord lived "lives of quiet desperation" in their efforts to cope with worldly demands or difficulties, and he suggested that they would be happier if they turned their attention to transcendental verities. Alcott, looking at these same ordinary people, saw that much of their desperation came from a heavy burden of illness. While he did not always try to measure it numerically, he thought that this burden could be substantially cast off by devotion to a systematic program of hygiene and disease prevention. "In the present blaze of physiological light," he wrote, "we can . . . manufacture human health to an extent not formerly dreamed of." Specifics for accomplishing this were essentially the same as for Graham and included exercising and breathing fresh air, moderation in eating and drinking, tempering the passions, stopping abuses of the nervous system (see Chapter 7), and ending the gross overconsumption of medicines. True preventive medicine seemed to have the potential to bring about and hasten a real millennium of health and longevity.[28]

For Alcott, Graham, and other hygienists, preventive medicine was a responsibility of the family unit as well as of the physician or individual. Accordingly, they prepared detailed instructions for the conduct of domestic life. Essential in such instruction was counsel about sexual impulses and activities. They warned against masturbation, provided some factual information on sexual anatomy, physiology, and venereal disease, and generally tried to turn young peoples' attention away from sexual impulses. Most of them maintained that an excess of sexual intercourse was physiologically harmful, and some assumed that it was sinful to find pleasure in intercourse. However, not all hygienists pushed this extreme view. Alcott himself felt that intercourse should be limited to once a month, but he regarded this as a means of controlling and refining the appetites, not denying them. In his view, "the pleasures of love, no less than the strength of the orgasm, are enhanced by their infrequency."[29]

Tracts of the Fowlers capitalized on the fact that many, perhaps most, Jacksonians were more interested in their propensity for "amativeness" or "sexual and connubial love" than in any other phrenological traits. In his intimate "suggestions to lovers and the married," for instance, Orson Fowler assured the public that the moderate enjoyment of sex was natural in a holy and healthy marriage. Other radical health reformers such as Thomas Nichols, Mary Gove, and R. T. Trall eventually went still further by introducing information about contraceptives into their physiological tracts, information which permitted even greater enjoyment of sex. Such publications, of course, proved to be more than most of the arbiters of respectable society could stomach. Nevertheless, even without such endorsement, the tracts did feed a widespread antebellum desire to know more about sex, while also doing something to replace prurience with healthy sexual attitudes.[30]

Hygienists generally agreed that marriage rather than singleness was the healthiest and most natural state for humans. Marriage was desirable because of the need for children to fill up the empty spaces of growing America, and because of the need for their labor on farms or in businesses. In many areas early marriages continued to be encouraged by the relative ease of earning a living in a rich, free, and democratic country. Thomas Hersey thought that where these conditions existed, they tended to stimulate marriage by making sexual life full and uninhibited. This was far from being the case, he felt, in Europe with its Malthusian problems, as well as in such American locales as New Harmony, the Shaker colonies, or New England with their religious restrictions or cold intellectualism.[31]

The received medical opinion of the day also tended to encourage marriage by its belief that single people were more liable to become insane than the married. Reports of the new insane asylums seemed to confirm this, and Edward Jarvis, during the early forties, sent out over 250 questionnaires to correspondents around Massachusetts in order to obtain further information about the subject. His returns indicated, reassuringly, that there were four times as many single insane people in that state as there were married ones.[32]

Hygienists placed much emphasis on proper moral, intellectual, and physiological preparation for marriage. By the 1840s it had become commonplace for middle-class couples not only to examine each other's phrenological characters but to determine each other's physiological condition before marrying.[33]

In yet another aspect of private life, Jacksonian hygienists, like those of the previous generation, urged people to improve their habits of dress and personal cleanliness. Prominent among their targets were style-conscious ladies who persisted in wearing fashionable Parisian garments at the expense of their health. One writer ventured the idea that in any given winter, shocking numbers of deaths could be expected as a result of violations of the code of sensible dressing and living. Among the dead would be:

> 1500 young girls between 16 and 22, of satin shoes and gossamer stockings.
> 1200 of being out till 3 o'clock in the morning, after going to parties till 10 at night.
> 1000 of tight lacing and gossamer dresses.
> 8000 of bare necks, and throats frozen for want of covering.[34]

Another observer estimated that as many as 20,000 females committed suicide annually in the United States by wearing garments with tight lacing. Those who had been reading the Belgian scientist L. A. Quetelet, moreover, knew that tight lacing had been associated with the rise in stillbirths, and some of them called for equally careful quantitative studies of these matters in the United States.[35]

If actual statistics on the effects of ladies' dress habits on health were still virtually nonexistent, those reflecting the extent of Americans' bathing habits were equally scarce. Nevertheless, some broad guesses were made. Noah Webster had urged earlier that city dwellers could undoubtedly increase their longevity by more regular washing. In fact, he warned, even the most active "sanitary police" could do little against an outbreak of the yellow fever unless the citizens also adopted hygienic modes of living, and particularly gave more attention to bathing. The novelist Charles Brockden Brown concurred in this advice, and saw no prospect that his American contemporaries would be able to save themselves from yellow fever if doing so depended on washing. According to Brown in 1804, there was an "almost general disuse of the bath. Vast numbers pass through life, amidst all these heats, clothed in cloth, flannel and black fur hats and lying on a feather bed at night . . . and never allowing water to touch any part of them but their extremities, for a year together."[36]

The well-to-do, of course, increasingly traveled to mineral springs, where "taking of the waters" was sometimes external as well as internal. John Bell catered to this fashion by surveying the various mineral springs around the country, classifying them by type, and furnishing tabular analyses of their chemical composition.[37] In the bigger cities, washing was being encouraged among the lower classes by the appearance of an occasional public bath, and in Philadelphia, the habit became infinitely more convenient with the completion in 1801 of the municipal water system. Physicians and moralists urged people to use it. Some Philadelphia scientists, with the advantage of the new abundance of water and influenced by Count Rumford's experiments on steam, undertook studies of water in its various forms and in its human effects.[38]

While Thomson and his "steam doctors" disseminated the concept of bathing as a heroic therapy, physiologists and health reformers tried to popularize it in the interest of cleanliness and the prevention of disease. Observers noted that some increase in the demand for soap and water did result from these efforts, though few figures were available to document this progress. Throughout the antebellum period, it was conceded by everyone that much remained to be done along this line. With the exception of a few urban areas, the observations of Harriet Martineau in 1837 continued to apply until 1860: "In private houses, baths are a rarity. In steam boats, the accommodations for washing are limited in the extreme; and in all but first-rate hotels, the philosophy of personal cleanliness is certainly not understood."[39]

The rapid expansion of urban water systems during the 1840s and 1850s enormously increased the use of water. More buildings had indoor plumbing, and by midcentury some

Americans were even attaching boilers to their kitchen stoves in order to have warm baths at home. As late as 1849, even in Philadelphia, where some fifteen thousand houses had running water, only about one-quarter had been equipped with bathrooms, but American cities were nevertheless better in these respects than British.[40]

Quantification of Dietary Practices

The same social, moral, and scientific impulses which encouraged personal cleanliness, conservatism in dress, and other reforms pertaining to hygiene were also factors in bringing about a newly critical consideration of diet. But statistical analysis played a considerably larger role in the advocacy, criticism, and assessment of dietary practices than it did in advancing most of the other aspects of hygiene.

An ideal of temperance in eating went hand in hand, of course, with that of temperance in drinking. The optimistic Rush thought that he could see marked progress in Americans' achievement of both. And he taught his students the importance of prescribing for patients a simple diet to supplement the more drastic therapies of bloodletting and purging.[41]

By all accounts, however, even during Rush's lifetime, the ordinary citizen seems to have ignored such advice. Noah Webster claimed that "multitudes" of Americans were no more willing to "regulate their diet [than to] cleanse their persons and habitations."[42] In the 1830s James Fenimore Cooper found Americans' eating habits utterly revolting, and he listed them with Americans' ignorance of good music and fine art as indicators of a deficiency in civilization: "The Americans are the grossest feeders of any civilized nation known. As a nation, their food is heavy, coarse, ill-prepared and indigestible, while it is taken in the least artificial forms that cookery will allow. The predominance of grease in the American kitchen, coupled with the habits of hasty eating and of constant expectoration, are the causes of the diseases of the stomach so common in America." Contemporary physicians went on to complete Cooper's picture, describing the businessmen in this " 'go-head' nation" who bolted large noon meals "with a fearful rapidity" (five to ten minutes), scarcely pausing to chew, "for they seem to consider their teeth as quite unnecessary instruments."[43] Travelers found things particularly bad on the riverboats and in towns west of the Alleghenies, where the meat was often cooked in fat that "looked like boiler oil," and where one encountered an overall "grossness of cookery and of preparation and quantity that offends the delicate eye."[44]

Such excesses provided a convenient focus for health reformers, both regular and irregular. In one direction, inventors like the versatile Charles Willson Peale devoted much attention to designing improved containers for preserving fruits, devising better utensils for the preparation of food, and generally applying new technology to the business of food.[45] Much more conspicuous, however, was the flood of published works on diet and digestion, aimed at both medical and general audiences and filled with facts and figures. Among the foreign works, Andrew Combe's *Physiology of Digestion* was particularly popular and widely quoted, but American medical authors also began to command attention, none more than the army surgeon William Beaumont.

Beaumont had no more than a preceptor's training in medicine along with a dozen years of military medical experience when he took charge of the unique case of the wounded French Canadian fur trapper Alexis St. Martin in 1822, at Fort Mackinac, in Michigan Territory. Over the next several years, however, he became fully aware that the observations of the digestive processes that he undertook, thanks to an unhealed flap in St. Martin's stomach, represented an important body of "incontrovertible facts." Beaumont proceeded to compile them in the best spirit of Louis's new numerical method, faithfully recording day by day the records of his many experiments. He supplemented his observations with chemical and microscopical examinations and correlated them in many cases with external weather conditions and with temperatures of the interior of the stomach. Among the most useful features in Beaumont's published account were his tables of the time required to digest various articles of diet, tables that health reformers subsequently quoted extensively.[46] Nevertheless, as Edward Jarvis pointed out, Beaumont's account was not notable for its statistical characteristics. Granting its great significance, Jarvis concluded that its generalizations could never be accepted fully until there had been "many more experiments of the same sort, and those tried upon other constitutions and in other circumstances."[47]

A more quantitative approach to diet was evident in John Bell's *Regimen and Longevity*. In it Bell made a generous use of "statistical calculations" to help give his readers a better idea of the subject of diet. He included chiefly figures on food production and consumption in various countries throughout history, together with data on mortality that suggested the ultimate consequence of bad eating habits.[48]

Bell, like most writers in the regular medical profession, approached the subject of diet counseling moderation. More radical writers, however, began pursuing dietary reform to the extreme position of vegetarianism. Graham and Alcott, who led this movement, both attracted their followers, and both argued, sometimes from statistics, that their positions were the proper ones to assure maximum longevity.

Graham, in urging adoption of his simple "Pythagorean regimen," or something like it, emphasized diet as the key to substantial reversal of the current decline in life expectancy. He pointed to recently published figures on Quaker longevity in England and the United States which he claimed proved the benefits of "plainness and simplicity and temperance in human diet." He also cited a Russian church census which attributed many cases of longevity to the "coarse and scanty vegetable diet" of the people, many of whom lived to be older than one hundred. And he attempted to show, from the 1830 census data, that the longevity of certain segments of America's population was due to the "physiological rectitude" of their eating habits.[49]

Graham directed his arguments on hygiene primarily to the poor; he felt that until they could live rationally and adopt a healthful diet, democracy would be subverted by the power of the rich. "While the poor continue to expend their trifling incomes for the most penurious luxuries, they will remain ignorant, and thus neglect the only means calculated to elevate them to an equality with their opulent neighbors." However, the idea of limiting food and drink apparently had little appeal for people who were already deprived of much enjoyment of life to begin with. It was the well-born, the intellectuals, and some middle-class zealots who took most earnestly to Graham's spartan program for good health. Hence, neither the Grahamite boarding houses which sprang up in many communities nor the Graham table that was founded at Brook Farm were filled by the poor, but by the privileged.[50]

Testimonials to the virtues of such programs were abundant, but there were few specific figures pertaining to their results. To be sure, a vague report circulated that, during the cholera epidemic of 1832, New York Grahamites who followed the regimen faithfully "either wholly escaped the disease, or had it very lightly." By and large, though, the Grahamites admitted that few of them had sufficient discipline even to stick to the special regimens for very long, let alone keep records or analyze the results of the diet.[51]

William Alcott, who did both, saw that achievement of dietary reform in American society could not be hoped for overnight but required a permanent educational effort among the people as part of the total movement for better personal hygiene. Alcott was drawn into taking an extreme position on diet by the results of a questionnaire on the effects of vegetarianism that a Hartford physician, Milo L. North, had circulated in 1835. When North was unable to follow through on the project, Alcott used the information as the basis of a publication of his own in 1838. Starting with the replies of perhaps two dozen correspondents, he added statements on diet and digestion by writers ancient and modern, along with the results of his own experience, to build up a medical, economic, and religious case for eliminating meat from the diet. Alcott calculated that, at current rates, a family could obtain over 25 ounces of vegetable nutriment for what it cost to get 5 ounces of meat. He estimated that if all Americans switched to vegetable food, the same amount of land which was supporting 21,000,000 people in 1839 could support 66,000,000. And he argued that, in the coming millennium of social progress, animals would be so completely crowded out by the growth of the human population and machines that by necessity meat eating would have ceased. However, Christians should not wait for such a millennium, for vegetarian diet was a moral imperative that could not wait.

Alcott presented a few examples from recent American experience which indicated that both morbidity and mortality were reduced by use of a vegetable diet. The Orphan Asylum of Albany had reported that between 1830 and 1833, when meat was included in the meals, "four to six children were continually on the sick list," and there had been over thirty deaths. After the fall of 1833, however, when a vegetarian diet was adopted, "the nursery was soon entirely vacated . . . and for more than two years no case of sickness or death took place."[52] Equally interesting was the record of the two-hundred-odd members of the American Physiological Society, which Alcott had been instrumental in forming in 1837. In the two years of the existence of "this society of vegetable eaters," many of whom were said to be sickly when they joined, Alcott observed great numbers of recoveries. Moreover, only four deaths occurred, a figure that was "only half the usual proportion" for the United States, he thought.[53]

Along with meat eating, another cause of the widespread chronic illness, feebleness, and early old age as Alcott saw it— one of "the numerous tributaries that combine to form the mighty river of premature death"—was the excessive drinking of tea and coffee. In a separate work Alcott drew together the history, medical opinion, and statistics of the consumption of these two beverages, which he regarded as drugs nearly as poisonous to the body as alcohol. Government statistics on imports for the years 1821 to 1838 showed that consumers had spent $125,000,000 on tea, and more than that on coffee. Alcott urged his readers to keep in mind the great "constitutional debility" and "especially the mighty host of nervous diseases" which such expenditures implied for citizens of the United States, and he listed better ways for Americans to spend that money.[54]

Graham's and Alcott's statistics gave an authoritative ring to the basically moral arguments they were trying to make. Undoubtedly, they were persuasive to the many who already shared the same assumptions about society, morality, and hygiene. Health faddists in considerable numbers welcomed the works of both men, as did many botanical and other irregular physicians.[55] By contrast, most regular physicians lacked enthusiasm for the dietary ideas of Graham and Alcott, and they were little impressed by statistics on vegetarianism. One reviewer criticized Alcott for drawing conclusions about the medical value of a vegetarian diet from a paltry twenty-three testimonials.[56]

The specific statistics that most stirred up the regulars were those relating American longevity to diet. New England doctors were especially incensed, both by the allegations of gross eating habits in that region and by the imputed resulting decline in average longevity. Some of them rushed to the defense of the area with polemical assertions that New England "probably" had the best record in the world in this respect. Alcott readily replied that "probablies" were not facts, and he criticized the tendency to rely on estimates, a practice which often reduced even famous writers to "retailers of hearsay."

Forewarned, a number of other New England physicians came forward with more specific figures on the subject. These data tended to prove that traditional rural habits of eating meat in large amounts, far from being detrimental to health, actually promoted longevity. John Bertram of Townsend, Massachusetts, thought that if enough country doctors would compile similar statistics, their records might well settle the whole debate on diet quickly.[57]

Alcott, who had kept a register of the vital events of his town even before he became a physician, applauded this idea, because he was convinced that complete records would prove him right about the evils of eating flesh. In any case, raising the longevity question in the context of the vegetarian crusade undoubtedly had the desirable side-effect of helping create, in Massachusetts anyway, a receptive atmosphere for improved vital statistics registration.[58]

Undoubtedly the most careful scrutiny of claims for vegetarianism by a regular physician was the study conducted by Charles A. Lee of New York. In the early 1840s Lee undertook a pioneering survey of dietary practices in America. He first questioned the improvements in the health of children at the Albany Orphan Asylum who had followed a vegetarian diet.

There was no doubt, he found, that distinct improvements in health had occurred there, but a close look revealed no evidence that it was caused by the elimination of meat. Rather, Lee's facts indicated that the improvement resulted from a much broader hygienic reform undertaken by the management: an insistence upon cleanliness, bathing, and exercise in addition to the new simple diet, together with the move of the school from a poor location to a large and well-ventilated new building.

Lee subsequently gathered information about the food provided in American orphanages, almshouses, prisons, and insane asylums. In 1844 he published his data on some eighteen institutions, comparing American practices with English and urging further study. There was no question, he found, that inmates of such places enjoyed better health where an abundance and variety of food, including meat, was provided. He noted that in 1832 the residents of poorhouses where diets were inadequate suffered "extraordinary fatality" in the cholera epidemic that year, while those providing generous diets had "very slight" mortality. He also noted the decreased likelihood of rebellions in almshouses that supplied residents with adequate food. Boston's institution already had a good reputation for that. In fact, a story was being circulated in that city about an "Irish pauper . . . who wrote home, advising his relatives to come out to this establishment immediately, as they *had meat twice a week.*"

It was as clear to Lee as to any vegetarian that a good cook was more vital to such institutions than a physician. Moreover, if officials would only spend enough money to provide hearty meals instead of pale thin soups, he predicted that they would see "as great a transformation as any we read of in Ovid" in the health of the inmates. He urged other Americans of his generation to start inquiring into the nature and extent of such dietary influences. At the least, he thought, it was worth considering "whether a greater ratio of cures would not be performed, both in public institutions, as well as private practice, by attaching more importance to the *Materia Alimentaria,* and less to the *Materia Medica.*"[59]

Applied to all aspects of hygiene, not just to diet, the therapeutic skepticism which prompted this type of observation had the effect of reinforcing Americans' perception that they could get by with fewer physicians. It encouraged continuing experimentation with new forms of therapy. And it made the public—the patients—increasingly receptive to new medical sects that continued to come into existence.

6

Quantitative Dimensions of Medical Sectarianism

The growing influence of unorthodox medicine was profoundly disquieting to the regular medical profession, already dissatisfied by the continuing lack of firm medical knowledge about disease and the obvious futility of much of their therapeutic effort. Regulars were also distressed by the removal of the professional protection that state licensing laws had afforded them and were disillusioned by the large numbers of incompetents entering the orthodox ranks from the proliferating medical schools.[1]

By the 1840s medical dissenters had multiplied from a scattered few irregulars who could be tolerated or ignored to an impressive number of practitioners increasingly well organized in distinct medical sects. With such organizations and numbers they mounted ever more aggressive attacks and became progressively serious professional and economic threats to the regulars. For the organized sectarians, collections of statistics became increasingly essential tools in promoting and defending their particular concepts of therapy. For the regulars, the gathering of quantitative information was an indispensable part of their efforts to discredit and beat back the unorthodox. In both of the camps an immense amount of energy and quantitative effort was thus devoted to the desperate struggle for medical status, respectability, and survival.

Numerical Factors of Medical Competition

In almost every local orthodox medical society in the land after 1840, old-timers periodically cast a pall over the proceedings with their depressing jeremiads about the gloomy prospects of the regular profession. The proliferation of sects went on, they lamented, despite the best efforts of the regulars. In fact, the unorthodox increasingly flaunted their ideas through organized means, expanding their networks of societies, schools, and journals. The Thomsonian movement was no longer a unified threat, but its clusters of botanic offshoots were proving more noisy, combative, and worrisome than ever. Meanwhile homeopaths, eclectics, and hydropaths were showing themselves to be even more dangerous competitors.[2]

Leaders in orthodox medicine went out of their way to assure their colleagues that, for all the noise and fuss, the irregulars and sectarians remained distinct minorities. Just how numerous they were, however, was a matter of debate. Estimates or enumerations were made in many communities, but the figures obtained were seldom very accurate. Dan King, a Rhode Island regular, thought that throughout these decades, in the nation as a whole, the regulars still outnumbered all others by more than four to one. In 1858, he posited about 31,000 regulars and perhaps 3,500 more or less organized irregulars, which he broke down as follows:

Homeopaths	1,000
Hydropaths	400
Female Physicians	300
Eclectics	800
Botanics	600
Chrono-Thermalists	300

He also estimated that there were about 3,000 or 4,000 Indian doctors, quacks treating cancer and venereal diseases, mesmerists, bone setters, nostrum vendors, and so on.[3]

Many of these irregulars were itinerants, but it was nevertheless easy for almost anyone who was a good talker and mixer to establish himself in permanent practice somewhere. All that was necessary, it seemed, was to post a sign in front of one's door with the word *Doctor* prominently displayed, and then "unite with the masons, the odd fellows, or sons of temperance . . . and, as a natural consequence, he is immediately patronized and puffed by some of his brethren!"[4]

To a beleaguered regular physician, there was little comfort in the figures showing that the orthodox were predominant, nor did it matter much that orthodox medicine itself was growing vigorously, so long as unorthodox medicine was perceived as a tangible threat to his livelihood. The irregulars, of course, perceived the orthodox as similar threats, so the medical world of the midnineteenth century became increasingly volatile and competitive, with the rhetoric and invective of the combatants tending to obscure the efforts at scientific advance. One outsider thought that even the most dogmatic contemporary religious sects were no more hostile toward each other than were the various segments of the medical community: "The eclectic, in the eyes of the regular school physicians, is a *quack;* and the homeopath, in the eyes of both the eclectic and the regular, is an *empiric;* while the homeopath, in turn, looks upon the allopath as a bigot." The latter, of course, regarded all the others as "worse than sinners."[5]

The ongoing doctrinal differences among the regulars—the contagionists versus the miasmatists, the therapeutic activists versus believers in the power of nature, for example—clearly enabled the other medical sects to flourish. Nevertheless, despite their internal differences, the regulars tended to agree with each other about the sectarian menace. At the same time, orthodox editors recognized that accounts of the activities of the irregulars were eminently newsworthy, and sought particulars from all sections of the country about the nature and magnitude of the sectarian threat. Allopaths of Rochester, New York, a city of over forty thousand in 1854, reported their ongoing conflict with large numbers of "foreign and native quacks, of all denominations and both sexes, [with individuals who endorsed] steam, water, caloric, magnetism, 'infinitesimal', and divers other species" of therapy. The five regular physicians of Ottawa, Illinois, in 1853 had to compete for patients with fifteen other practitioners representing "nearly every system of error" and making the place of the regulars "very unpleasant" as well as highly insecure in that town of four thousand inhabitants.[6]

Although the public at times seemed oblivious to the differences among the various kinds of practitioners, the regulars tried to make them clear—at least the difference between themselves and all the others, whom they regarded as incompetent. Like Samuel Cartwright, Worthington Hooker claimed that this matter of medical competence could be readily measured by the results of one's medical practice:

1. He [the skillful regular physician] has a less number of fatal cases in proportion to the whole number that come under treatment.
2. He has a less number of bad cases . . .
3. His patients have commonly a shorter sickness.
4. They are in a better condition after they have recovered.
5. He has a less number of patients, and a smaller amount of sickness, in the same number of families.[7]

Hooker's commonsense criteria might well have been useful both to the medical profession and to the ordinary citizen in distinguishing between the best and the worst of antebellum physicians, but unfortunately, the data for making such distinctions were not readily available, so there was no effective way to judge between an outright quack and a skillful practitioner. The same was true in trying to decide about home medications. Patent medicine manufacturers were well aware that claims of large numbers of cures in their advertising would lead consumers to purchase a particular drug. The manufacturer of Sarsaparilla, for example, claimed during the 1840s that the mixture had sold so widely that, in only two years, thirty-five thousand cases of severe diseases had been cured. It remains a question whether the readers of that day were more impressed by the numbers of alleged cures or by the volume of business.[8]

However figures were used in drug advertising, few if any regular physicians seem to have used statistics to evaluate the medical sects. For one thing, obtaining unbiased numerical data from the sects was virtually impossible, and, too, many regulars did not take some of the sects seriously enough to consider making laborious numerical assessments.

Even when statistics were available, though, they posed basic problems, since every sect seemed to have a "predilection to flattering statistics." New Orleans's Bennet Dowler noted that such practices tended to make a mockery of serious attempts to use statistical method: "Statistical argument, however decisive in itself, is virtually neutralized by the significant fact that every mode of treatment from the heroic to the infinitesimal claims its protection and takes shelter under its autocratic sanction."[9] Nevertheless, despite such misuses, sectarian statistical practices are well worth examining in detail. Of the principal organized sects, none valued numeration more than the homeopaths.

Homeopathy

During the 1840s and 1850s homeopathy emerged in the United States as the most durable, and to some the most appealing, of the unorthodox medical sects. Unlike Thomsonianism, homeopathy was a medical import, one that came to the attention of Americans at the very moment that a few of the orthodox were starting to learn the numerical method from Louis. For many educated people, homeopathy seemed to provide an especially promising rational alternative to the existing medical systems. The fact that it was even more Baconian in its approach than the numerical method did nothing to lessen its appeal, especially for physicians.

Although by 1810 the German physician Samuel Hahnemann had published his *Organon der rationelle Heilkunde*, the work in which he first described homeopathy, American physicians did not begin to embrace the system until the mid-1820s. For a number of years after that, only a handful of medical men on this side of the Atlantic pretended to understand the main points of homeopathy, either the doctrine of "like cures like"—the treatment of diseases by drugs producing similar symptoms—or the rationale of relying for cures upon extremely tiny doses of such drugs, or "infinitesimals," as they were called. However, by the midthirties a few regulars had converted to homeopathy, and immigrant ships from Europe were bringing a number of German physicians who soon converted or who were already firm believers in Hahnemann's methods. Some of the early arrivals had qualifications not only in medicine but in the physical sciences, mathematics, or other fields. One homeopath, William Wesselhoft, during his medical studies at Jena had worked as a "clerk of the weather" for Goethe.[10]

Concentrated at first mainly in the East, homeopaths gradually established their own journals and societies, organized medical schools, set up their own drugstores, and established clinics and dispensaries. In short, homeopathy became a distinct medical enterprise. From the beginning its advocates worked to undermine and overthrow the therapeutic practices of conventional physicians—especially bleeding patients and administering drugs. They and their followers were so successful in this that by 1860 many thought the entire American medical world was teetering "on the brink of Homoeopathy."[11]

One of the aspects of homeopathy that Jacksonian Americans could readily understand, and one that brought it a certain number of converts from the regular medical profession, was its emphasis on the systematic pursuit of facts and on the use of orderly methods. Hahnemann attached somewhat the same importance to precision of observation and exactness in gathering data that Louis did. In fact, Hahnemann had urged upon his followers the necessity "of addicting themselves to close thinking, by the study of Mathematics, of qualifying themselves for minute observation by the study of natural history."[12] The intellectual requirements of homeopathy thus demanded as much motivation of its devotees as Louis's method required of the young regular physicians who went to Paris. The two groups of practitioners were equally convinced that theirs was the medicine of the future, because theirs was based on a scientific methodology, on true Baconian principles of inductive philosophy. Indeed, Baconianism was so pronounced in homeopathy that the regulars' efforts to achieve orderliness and numerical precision seemed bumbling and feeble by comparison.

William H. Holcombe boasted of homeopathy's success in bringing "Mathematics into the hitherto uncertain field of Therapeutics," and he saw the system eventually making use of physiology, psychology, chemistry, and physics as well.[13] The more typical midnineteenth-century homeopathic practitioner, however, had a more limited concept of it. For him, homeopathy stood on two supports. One was its original set

of basic principles; the other, the record of successful application of these principles as expressed by "the stern language
of statistical tables." Homeopathic spokesmen knew that generous compilations of such data could "give confidence to the
practitioner as well as to the intelligent laymen, while they are
unanswerable arguments to our opponents."[14]

Yet, like the "allopaths" (the practitioners of orthodox,
mainline medicine), the homeopaths had to be continually
urged to maintain records of their practical experience. Homeopathic editors regularly solicited and published such statistics. Members of local homeopathic medical societies,
meanwhile, busied themselves with collecting data on the diseases of their regions. Eventually, new homeopathic infirmaries and hospitals also carefully amassed and published the
statistical records of their patients.[15] And, in the mid-1860s
the American Institute of Homoeopathy even established a
Bureau of Organization, Registration, and Statistics, for scientific as well as administrative purposes.[16]

Constantin Hering, an immigrant physician who was one
of the ablest and most articulate of the early homeopaths in
this country, made clear the basic importance of good record-
keeping in homeopathic practice. He noted that for homeopaths the examination of the patient was a crucial procedure,
one that was still often done only superficially by orthodox
physicians. It required at every stage, Hering emphasized, the
most meticulous gathering and recording of data: listening to
the patient attentively, writing down his statements as fully as
possible, eliciting further information by a form of Socratic
inquiry, and classifying or arranging the whole to form a rational basis for making a diagnosis or prescription. This done,
the homeopathic physician then went on to keep other detailed
records of the treatment adopted, the drugs prescribed, and
the effects achieved. In this way the practitioner built up his
own personal store of information about disease. As one new
convert observed, "What a mass of facts may not be acquired,
in a few years, by a close observer?"[17]

Among the observations made by the homeopathic practitioner, none were more important than his "Provings" of drug
effects. Hahnemann had worked out the original provings—

tests to determine the symptoms produced by drugs of various strengths—on his own body. His followers were not obliged to take his findings on his authority; all were urged to learn the effect of drugs by repeating one or more provings to their own satisfaction. American physicians, like their European counterparts, did this as part of their training in homeopathy. In fact, one of the first organizational steps taken by Pennsylvania's early homeopaths was to establish a Provers' Union.[18]

Whatever else it may have been, the proving was an important part of the intellectual attraction of homeopathy. In carrying out the provings, the new homeopathic physician seems to have considered himself far more of a scientist than he had been while empirically prescribing or blindly mixing up allopathic drugs. Since this was a highly personal form of scientific inquiry, moreover, those engaged in it could hardly have felt inhibited by the current prejudices against research as a career.

Homeopathic therapeutics required ongoing scrutiny of a number of other quantitative procedures: there had to be constant reevaluations of the different drug "potencies" being utilized, recalculations of dilutions, recalculations of the proper lengths of time for pulverizing substances, and refiguring the numbers of shakes to be administered during the preparation of liquid mixtures. To facilitate the huge numbers of shakes necessary for manufacturing some of the higher potencies of drugs, homeopaths experimented with various mechanical devices. Constantin Hering used a sawmill engine to shake some of his remedies hour after hour, and another homeopath, B. Fincke, used a steel spring for mixing some remedies and exposed others "to the motion of the railroad and ship conveyance for about seventy days."[19] Some Americans even suggested means for standardizing the mathematical notation used to express homeopathic quantities.[20]

The most crucial data for the homeopaths, however, were statistics reflecting the movement's therapeutic performance vis-à-vis the allopaths. The first generation or so of American homeopaths had, frustratingly, little specific local ammunition to back up their claims of superiority. A few statistics were available from the various governmental testing of homeopathy which had been conducted in Europe during the 1820s and 1830s. Homeopaths claimed that these were favorable to their cause, though allopaths viewed them quite differently. In any case, the earliest reports from native homeopathic practitioners were eagerly welcomed in the United States.[21]

In one report, a Michigan physician claimed to have had complete success in the homeopathic treatment of 200 cases of "Western" fevers, especially remittents and intermittents, while the allopathic treatment in his community normally resulted in the patient's being "puked and purged straight to death." In rural Mississippi, William H. Holcombe and F. A. W. Davis reported that they treated 1,016 cases of yellow fever homeopathically during the years 1853–1855 with a mortality of 55, or 5.4 percent, while according to their estimates, the allopaths in the region were losing between 20 and 25 percent. Homeopaths in Lowell, Massachusetts, reported that 1 out of every 12 persons treated for dysentery by regular physicians in an 1847 epidemic had died, while of the 156 treated homeopathically, only 2 had died.[22]

The data from individual practices were particularly important because, prior to the Civil War, few facts about homeopathy could be obtained from hospital populations. Most hospitals, almshouses, and other institutions in this country barred sectarian practitioners from their wards. In the 1840s and 1850s, however, a few institutions did allow homeopaths to practice in them.

At New York City's Protestant Half-Orphan Asylum, the homeopath Clark Wright became sole attending physician in 1842. Wright immediately introduced a strictly homeopathic therapy for the many cases of ophthalmia, and he followed this by using infinitesimals for all diseases. Over a period of ten years, he and his homeopathic successor claimed an enviable rate of cure at the asylum. It was, they asserted, over three times the success achieved in the city's six other orphanages, all of which were under the care of allopathic physicians. New York homeopaths used these figures to urge that all asylums in the state be required by law to file full statistical reports so that the public might appreciate the advantages offered by homeopathy in saving lives.[23]

During the 1850s homeopathic physicians reported comparable triumphs over the allopaths at orphan asylums in Philadelphia, state prisons in Michigan and New York, the Five Points House of Industry in New York, and other institutions. Claims of reduced death rates from homeopathic treatment at such institutions were sometimes reinforced by figures on the savings of time lost due to sickness and of the money spent for medicines.[24] Unfortunately, however, there were not yet sufficient institutional experiences with homeopathy to provide the sect with much statistical ammunition against the allopaths during the country's second great wave of cholera. Consequently, its followers achieved their most telling points by criticizing the records of the allopaths.

European homeopathy had gained its greatest early impetus from its reputed successes during the cholera epidemic of 1831–1832. When the disease again threatened the United States in 1849, American homeopaths were confident that they stood to benefit similarly. If nothing else, the appalling impotence of American allopaths in the earlier outbreak seemed to make infinitesimal therapy, as homeopathic treatment was often called, an attractive alternative. New York homeopaths thus made all they could of the 1832 record, which reportedly showed that some 2,000 of the 5,200 New Yorkers who had come down with cholera had succumbed to the disease. In 1849, cholera mortality in New York's allopathic institutions reached a comparable level, and in Philadelphia, Cincinnati, Boston, and other cities, homeopaths published figures which showed regular medicine in a very unfavorable light. As the homeopath Benjamin Joslin put it, his colleagues felt that in that situation they had only to sit back and "let statistics answer this question" of which medical system was preferable.[25]

Joslin was convinced that much of the allopaths' helplessness with cholera arose as a result of their "not viewing the subject in a mathematical point of view, that is, in its relation to the science of quantity." He saw little likelihood that the situation would improve, since as a group the American regulars were characterized by a "deficiency of mathematical training."[26] Moreover, while the regulars had only a sprinkling of registry offices and health departments to produce official statistics on cholera, the homeopaths felt that they could draw upon almost every one of their individual practitioners as a potential contributor of relevant data.

As the 1849 epidemic pushed inexorably across the country, therefore, individual homeopaths did indeed rush into print accounts of their treatments and successes. They reported death rates of from 1 to 8 percent among their own patients while stressing that the death rates of patients treated by the inept allopaths in their localities ranged from 20 to as high as 70 percent.[27] Delegates to the annual meeting of the American Institute of Homoeopathy in the summer of 1849 saw the cholera statistics almost as a manifestation of God's providence, designed "to open the eyes of the people . . . in favor of our system."[28]

Allopathic Views of Homeopathic Numbers

The cholera epidemic, of course, was only one factor helping to spread homeopathy. Others were the fragmented state of the regulars, the low prestige of traditional physicians, and the appeal of a medical system that encouraged less rather than more use of drugs. The resulting growth of homeopathy was so pervasive and menacing that the regular profession had no choice but to greatly intensify its arguments against and resistance to the movement.

The regular medical profession as a body originally paid little attention to a form of therapy that was practiced mainly among the foreign-born elements of the population (though a few native physicians were curious about it and not unsympathetic to it). In fact, the German origins and character of homeopathy, together with the frequent involvement of its partisans in Swedenborgianism and other foreign "isms," were sufficient to turn away a good many of the conservatives. When ordinary physicians did notice the sect, it was mainly to ridicule the presumed absurdities of the theory of infinitesimals without devoting much time to actually examining it. Of course, ridicule was one of the penalties paid in antebellum America for holding unorthodox medical views. Ardent homeopaths rationalized that such a response to their therapy was at least better than being totally ignored. One enthusiast even suggested that contempt was an encouraging sign—the predictable second stage of a three-stage ladder of progress which led toward ensured acceptance.[29] William H. Holcombe of Cincinnati assured his fellow homeopaths that the regulars' "vituperation is proportioned to their ignorance," and he was certain in 1856 that the success of his sect had become "one of the fixed facts in the public mind, which neither allopathic statistics nor sarcasm can uproot."[30]

Certain leading regulars, to be sure, had some limited praise for a therapeutic system which had the virtue of reflecting their own cautious approach to drugs and reliance upon nature. James Jackson, Sr., Jacob Bigelow, Oliver Wendell Holmes, and Erasmus N. Fenner, among others, went out of their way to commend homeopathy during these years for having rejected the heroic therapeutics of the day. But these favorable notices were in turn subjected to such harsh criticism by conservative regulars that Bigelow, Holmes, and the other liberals were forced to take strong positions against other aspects of homeopathy in order to protect their professional standing.[31]

Allopaths deeply resented the posture of superiority that many homeopaths assumed. As professionals, they opposed the exclusivity of homeopathy's doctrines and were convinced that its materia medica of infinitesimals was no more than a collection of placebos. Along with these qualitative objections, moreover, the allopaths took strong exception to some of the quantitative aspects of homeopathy.

From the beginning, the arithmetic of infinitesimal fractions was a primary barrier to the comprehension and acceptance of homeopathy by conventional physicians. Unaccustomed to such mathematical exertions, even the most sympathetic of the regulars complained of "mental fatigue" in their efforts to understand such complicated calculations, though homeopaths insisted that using them was necessary if medicine was to become a true science. Certainly, the exertion was no more than was required in keeping up with "the wonderful calculations of astronomy."[32] But that was just the problem. As a result, American regulars who had any bent for mathematics used it less in trying to understand the new practice than in attempting to establish the absurdity of its claims.

In this spirit, Holmes in 1842, Hooker in 1851, and a veritable host of unbelieving regulars filled pages of their arguments against homeopathy with calculations. They confidently believed that the data showed the inherent preposterousness of infinitesimal "logic." Actually, it took little mathematics to be able to heap ridicule on the arithmetic of "dynamization," that vitalistic process of imparting a spiritual element to drugs. Regular physicians were from the outset both mystified and amused by the claim that the highly attenuated substances of the homeopaths, far from losing strength when diluted, would actually gain some mysterious power in the process of trituration (for solids) or shaking (for liquids). Merely a mention of the solemn homeopathic concern with how many shakes to perform at each dilution was sufficient to convulse meetings of almost any orthodox medical society with laughter. Hahnemann's cautious advocacy of two powerful shakes of the arm seemed odd enough, but the thought that some American practitioners would actually deliver up to a million and a half shakes to obtain a high potency was beyond belief. "Who would not pity poor Joenichen's arm?" Why would not a few great shakes do the job as well as many smaller ones?[33]

Even more entertaining for the regulars were the mathematical exercises which calculated the volumes of fluids or solids necessary to obtain the various homeopathic dilutions. These dilutions, as Hooker pointed out, formed parts of a "towering arithmetic" of therapeutics. The system used not just such trifling and comprehensible quantities as hundredths or thousandths but dealt often in millionths, billionths, and trillionths. It was obvious that the arithmetic of homeopathy went "beyond all chemistry—no test can reach its higher dilutions." Regulars who were accustomed to prescribing calomel by the tablespoon thus roared at the suggestion that one grain of homeopathic oyster shell properly diluted and used as a medicine would suffice the entire world for all time. They were equally amused to note that merely the twelfth dilution of a homeopathic liquid would require an amount of alcohol that would fill five hundred lakes the size of Lake Superior.[34]

However much they relished the humor in such figures, the regulars were not at all amused by the statistics the homeopaths produced, particularly published data on homeopathic practice. Regulars readily concluded that such data were incomplete, misleading, and sometimes deliberately distorted. Holmes, as early as 1842, warned against the sect's "sweeping statistical documents," which, on close inquiry, proved seldom to cover "anything but successful cases." A striking instance of this was uncovered in 1848 by the black physician James McCune Smith. Smith, finding that homeopathic health claims for New York's Protestant Half-Orphan Asylum had a "suspicious look," made a personal investigation of the asylum's records. He found that, owing to various discrepancies, the claimed death rate of 1 in 144 was really more than 1 in 90. In addition, however, he discovered that large numbers of children had been unaccountably released from the asylum during the five-year period before the report was made. Although Smith found no firm evidence, the dismissals suggested to him the existence of "a custom of quietly thrusting away from this charity the very sick children, in order that they may die elsewhere." They implied a cold-blooded policy "to magnify Homoeopathy by clumsily contrived statistics."[35]

Regulars further resented the "fact that the homoeopaths seldom or never perform any serious or hazardous surgical operation—admit no moribund or desperate cases, etc.; but that their patients generally belong to that class of society among whom are generally to be found the nervous or hypochondriac, etc." Since the serious and hopeless cases tended to be left to the regulars, mortality reports understandably often favored the homeopaths. The allopaths had a hard time putting the significance of this discrepancy across, even to the presumedly intelligent classes. Indeed, as Bennet Dowler lamented, "the public tacitly seem to credit the flattering—O! most self-flattering reports of homoeopathists."[36]

Worthington Hooker held that the value of therapeutic statistics depended on the impartial character of the observer and on his consideration of all relevant variables. In both respects he found the statistics of homeopathy wanting. Figures produced by the sect could not be believed since its members were committed to preconceived conclusions of an exclusive theory. During cholera epidemics, moreover, homeopaths rarely attempted to differentiate between their many mild cases of diarrhea and the serious cases of cholera; as a result, they had great stories of success to pass on to the public. Other kinds of homeopathic statistics were equally unreliable, Hooker thought, for they were presented "too much in the advertising style of quackery."[37]

Homeopaths regarded the regulars as being highly presumptuous in attacking infinitesimal statistics at a time when allopathic data held so many "uncertainties, fallacies, inaccuracies, and speculations." They claimed, with some justification, that American regulars were in no position to criticize homeopathy because they had never tested it on its own terms; indeed, few if any orthodox physicians were felt to have the scientific background necessary to perform such tests. Holmes, however, thought that Gabriel Andral's negative statistical conclusions about homeopathy, announced in Paris in 1835, were good enough for him, and virtually every American allopath of the forties and fifties concluded the same thing. Elisha Bartlett, concurring, felt that homeopathy's utter failure to produce convincing statistics rendered any further tests unnecessary: "So far as I know, . . . homoeopathy has, in no single instance, on a scale of sufficient magnitude to be of any value, complied with the conditions which are absolutely necessary to ascertain the actual and comparative efficacy of its methods of treating disease."[38]

While orthodox leaders thus failed to take the claims of homeopathy very seriously, laymen comprehended fully that at least it offered a therapy that did no harm to the patient. Besides, as members of a generation that thrilled to the marvels of a P. T. Barnum and to the exploits of a Davy Crockett, they were as ready to try homeopathy as any other novelty. Antebellum Americans had become expert samplers. They sampled economic novelties; they dipped into a variety of philosophies; they tried religions of every hue; they copied the great architectures of history so assiduously that structures in neo-Gothic, Greek Revival, Tuscan, and Egyptian styles sprang up in every corner of the American landscape. It was, in short, an eclectic age. The ordinary person's infinite capacity for medical variety was one further manifestation of this trend.

It was only natural that some practitioners would actually pick and choose from the many medical concepts floating about and fit portions of all of them into the framework of still another sect, eclecticism.

Statistical Ingredients of Eclecticism

Like so many other individuals, those who became eclectics had weighed orthodox medicine and found it badly wanting. However, unlike some sectarians, they did not totally reject it. The central eclectic, or "Reform," solution was not different from rational medicine, homeopathy, or some of the other therapeutic systems in trying to get away from the heroic therapies used in the early part of the century. It was close to Thomsonianism in focusing heavily on native medicinal plants for their materia medica in preference to the regulars' potent metallic drugs. Yet the eclectic felt relatively little affinity for the untutored Thomsonian practitioner. The crude Thomsonian system was built around a handful of the plant remedies used in folk experience, while eclecticism was shaped much more substantially by the systematic findings of formal medical botany. Ultimately, moreover, eclecticism found a far greater place for the systematic use of statistics than Thomsonianism ever did.

The eclectics drew in a general way on the scientifically arranged botanical compilations of Manasseh Cutler and Henry Muhlenberg and on the beautifully illustrated tomes of Benjamin Barton and Jacob Bigelow. Much more specifically, however, they relied upon the practical medical botany of the colorful wanderer, Constantine Rafinesque. Rafinesque's manual was chiefly significant, perhaps, because, unlike the luxurious works of Barton and Bigelow, it was economically within the reach of Jacksonian Americans. While it may have compromised on the quality of its artistic renderings, it did include as many descriptions of native plants as its predecessors, as well as lists of many plants that could be used as substitutes in medical preparations. Rafinesque's full descriptions of medicinal plants thus provided a convenient guide for people who became eclectics. In his passion for nomenclature, moreover, Rafinesque neatly arranged the kinds of medical practitioners in America into three classes, each subdivided into "species." To one of the species he deemed most enlightened he gave the name "Eclectic."[39]

Although Rafinesque had no intention of fostering another new sect, that is essentially what happened. By the 1830s the pursuit and application of medical botany was falling into disrepute among the regulars as it became progressively more tainted by its connection with the unlettered Thomson. Rafinesque himself subsequently contributed to orthodox suspicions of plant medicines when he not only produced and marketed secret botanical formulas but, though not a physician, prescribed them for medical conditions.[40] Antebellum regulars thus found little difference between the eclectics and the Thomsonians and tended all too often to leave medical botany to them and other practioners of unorthodox medicine. As long as Thomsonianism flourished, moreover, eclecticism developed slowly, and it achieved some prominence only after the thirties.

Wooster Beach was a proponent of eclecticism, or "Reform," even before Rafinesque publicized the term. Before 1830 Beach made attempts in New York to institutionalize eclecticism by establishing an infirmary, a medical school, and a medical society, though these proved both too local and premature, since the concept was not yet well known. In the mid-1830s, however, following the publication of Beach's "Reform" treatise, *The American Practice of Medicine*, eclecticism began to be noticed. By then, observers could report as many as two hundred trained physicians for the movement, and probably an equal number of untrained individuals. Progress became even more visible during the forties and fifties with the emergence of a variety of Reform medical schools, journals, and societies, including a national organization. Ohio, which became the most active center of the sect, boasted some three hundred eclectic medical students in 1852, and another hundred or so were studying in upper New York and New England schools.[41] Many of the schools and journals, however, were short-lived. Some, while they existed, were scenes of violent disputes between the eclectics and others, especially members of the botanical sects, and occasionally even the mathematics used in Reform medicine became the subject of vituperative controversy.[42] Partly because of these disputes, many practitioners of eclecticism defected during the late 1850s, leaving the future of the sect very much in doubt at the beginning of the Civil War. Quite remarkably, however, only a few years later, eclecticism revived strikingly, and in the postwar decades it became the strongest irregular medical sect after homeopathy.[43]

The substantial institutionalizing of eclecticism led to the usual applications of statistics in propaganda. Eclectic editors took pains to find and publish data on the evils of over-drugging; one produced figures claiming that in Cincinnati alone, the sales of calomel had amounted to an average of seventeen tons annually between 1842 and 1845. Most of the editors, however, acting as mouthpieces of particular eclectic medical schools, were more concerned with the special status of those schools and with data that presented them favorably. *The Eclectic Medical Journal* understandably made much in the mid-fifties of the enrollment of Cincinnati's Eclectic Medical Institute: in only a few years the school had grown larger than any regular medical school either in Cincinnati or in Kentucky.[44]

Eclectics recognized the need to obtain greater support from the public before their movement could hope to expand very much. Considered essential for this process was "the accumulation of statistics proving the superiority of Eclectic practice, and showing its results in each disease, as well as the results of new agents and methods of treatment." The National Eclectic Medical Association thus regarded statistics as one of the seven fundamental branches of medical science, and established a committee under Joseph R. Buchanan to encourage development of statistical skills in its members. The principal activities of the Committee appear to have been its periodic appeals to society members to keep standard routine records of their private practices and to report them annually in the eclectic journals. Such appeals were echoed in local societies.[45]

Only a scattering of practitioners seem to have responded to these calls, or at least to have had their statistics published, but they reported striking results. In a sample year Dr. William King claimed no deaths in 82 cases of 6 major diseases; T. J. Wright had only 4 in some 200 cases of all kinds; and T. V.

Morrow, editor of the *Eclectic Medical Journal,* claimed nine deaths in 840 cases. During the cholera of 1849, eclectics rejoiced when orthodox and eclectic practice was subjected to a public "test of facts and figures," and the regulars seemed to have been less successful. In Cincinnati, the conclusion was that "the Eclectic physician saves twelve cases out of fourteen, while the old school physician saves eight or ten out of twenty-four."[46]

At a different level, a few individuals considered statistics not merely a tool for propaganda but the means needed to make eclecticism a precise science. Among them, Victor J. Fourgeaud of Saint Louis regarded the method of Bacon and Louis as one of the desirable elements of orthodox medicine that had to be retained. Only the patient accumulation of facts and their numerical comparison with the experience of the past, he thought, could hope to "establish Eclecticism on an imperishable basis." Hiram Cox painted a comparable picture of diligent eclectics systematically gathering facts to build up a body of general principles for their practice. In the process, he said, "we discard speculating theories, and employ statistics, which give the experience of every age of the profession." He saw eclectic therapy thus gradually becoming an effective amalgam of botanic, homeopathic, allopathic, and other remedies or principles. "We profess to treat diseases upon Botanical principles, and yet we do not entirely abandon minerals . . . We differ as widely from the Old School, as we do from the . . . steamers, yet use many of the remedial agents that are used by both." As Joseph Buchanan concluded, how absurd it was for a medical man to be labeled as a follower of one belief and to thereby limit one's therapeutic options. "American Eclecticism repudiates all such restrictions and endeavors to reach the quickest, best and surest cure for the patient, no matter whether it be by the Antipathic, Homoeopathic, or Allopathic law, or by no law at all that human philosophy has yet discovered."[47]

Some in the eclectic community thus were interested in a modest midcentury quantitative test for inoculation for rubeola. John E. M'Geer was the eclectic physician to two Catholic orphanages in Chicago at the time of a measles outbreak in

1851. Among the 29 infected girls in one of the institutions, M'Geer inoculated 15. All 15 recovered, but 5 of the 14 not inoculated died of the disease. In the boys' orphanage, where there were no inoculations, 6 patients out of 23 died.[48]

As it turned out, this investigation was exceptional; significant statistical clinical studies by eclectics before 1860 were few. The numerical method remained in theory one of the desirable legacies of orthodoxy to Reform, yet eclecticism had neither the hospitals needed to take advantage of the method nor a sufficient corps of trained practitioners who also had an aptitude for mathematics.

Hydropathy

Shortcomings in mathematics were even more conspicuous among individuals who practiced hydropathy, a therapy stressing intensive applications of water, internally and externally. Like the founder of their sect, Vincent Priessnitz, most hydropaths did not (at least at first) pretend to be learned, yet, unencumbered by such an incubus, they succeeded far better than the eclectics in propagating their therapy in the antebellum United States.

Americans discovered Priessnitz's water-cure establishment in Graefenberg, Austria, in increasing numbers during the late 1830s and early 1840s. They found few comforts to mitigate Priessnitz's strenuous therapy—he made them "work for health"—but some stayed with his regimen and some felt better after going through it. By 1844 an estimated eight thousand to ten thousand people from fifteen countries had taken the cure at Graefenberg.[49] These individuals were bound together, it seemed, by "the same horror of the mineral poisons." In 1844 Priessnitz was reported by one follower, Henry Wright, to be a man of extraordinary simplicity who was quiet but firm. Even after two decades of experience with his therapy he kept no records, had "no theory at all," and was regarded as being "incapable of giving a written account of his system."[50]

Wright hoped that the water-cure concept would soon spread to the United States. He confided to William Lloyd Garrison his dream of the country's being ultimately transformed by the concept into health, of cities and villages with no sick people, of doctors unemployed and praying for patients. Other early enthusiasts pointed out that the entire American economy and environment might be transformed by the water cure. The realization that water, which was abundant, was the only medication needed by the citizen would bring an end to Americans' hectic pursuit of commercial drugs. "Mankind will not then be under the necessity of ransacking all nature for medicine to cure."[51]

The water cure did come to the United States in the mid-forties and spread with great rapidity. By 1847 some fifty practitioners were reported in nine Eastern states, and twenty-eight water-cure establishments had been built, rambling hotel-like structures with varying facilities and advantages, and usually differing somewhat from the existing spas and health resorts. Ten years later, as the craze moved westward, the numbers of water-cure practitioners and establishments had both roughly doubled. Despite their frequently aired complaint that the United States was a "doctor-ridden land," hydropaths somehow found sufficient justification for launching a half dozen medical colleges during the fifties, though only R. T. Trall's New York Hydropathic School lasted more than two or three years. Apparently hydropaths felt little need to exchange professional ideas with their fellow practitioners, for organized societies were also rare. The American Hygienic and Hydropathic Association, which held meetings in New York between 1849 and 1851, in fact was perhaps the only one of its kind before the Civil War.

Hydropaths did produce some periodical literature, but this was mostly in the form of ephemeral bulletins touting particular water-cure establishments. The one important exception was the *Water-Cure Journal*. Founded by Joel Shew in 1844, this journal originally had the same limited circulation of most other medical periodicals. However, in 1848 it was taken over by the publishers Fowlers and Wells, who turned it into a monthly; within two years its circulation was a reputed fifty thousand, one of the largest magazines in the country. In the process they helped broaden the scope of water cure and turned it into the preferred therapy of a large segment of the middle class.

The introduction of hydropathy could not have come at a more favorable time. It flowed into the country on the waves of enthusiasm for sanitation generated by the completion of the great waterworks for New York and Boston. The appeal of taking a cure at a scenic mountain establishment, as at Graefenberg, was by itself very considerable. However, the good water abundantly supplied by the new public water systems made hydropathy feasible for many more. Without leaving the city one could now relatively inexpensively get the same varied water treatments offered at the rural establishments: the baths and showers of every description, the wet sheets and sweating regimes, the douches, enemas, and injections.

The extraordinary rate at which the early practitioners poured water into and over their patients emphasized the fact that hydropathy was indeed based originally on a single form of therapy, and some continued to believe that water by itself was enough to cure disease and maintain health. Yet, almost from the beginning, many others recognized that the value of water as therapy could be enhanced by applying it along with other regimens of personal hygiene. In fact, the early health reform activities provided a very favorable environment for the introduction of hydropathy. And conversely, many of the Grahamites, phrenologists, temperance enthusiasts, and popular physiologists who were already attached to the simple menus and Spartan hygiene of the Graham boarding-houses found the water-cure hotel much to their liking, and subsequently they did much to expand its therapeutic scope. Well before 1848, therefore, most American spokesmen thought of hydropathy as a two-sided phenomenon. Resorting to the full rigors of water therapy was essential only for those already sick. For others, an almost equally strenuous program of hygiene was dictated to maintain good health. The true believer in water cure believed equally in the prevention of disease through careful attention to diet, drink, exercise, clothing, sleep, housing, and personal cleanliness.[52]

It was no accident that the hydropaths were among the country's "most efficient instrumentalities" in overcoming Americans' traditional aversion to taking regular baths. They urged people to build bathrooms in their houses. They urged hospital attendants to bathe their patients. When the cholera returned in 1849 they were prominent among those who urged that cities provide free baths and wash-houses to promote cleanliness among the poor.[53]

As disciples of Priessnitz and as reformers with very firm faith in their measures, hydropaths seem to have felt no great compulsion to keep records of their practice or to make it more precise. One practitioner, C. C. Schieferdecker, paid attention to such factors as "the form of the bath, of the time, of the temperature of the water, of the repetition, of the change," and so on, as well as combining different kinds of baths in treating his patients. However, he made no attempt to calculate with any exactitude which factors or combinations of factors were appropriate for the treatment of specific diseases or conditions. As for results, he was content simply to assert that he had "not lost one patient out of the hundreds of desperate cases I have treated in this country with water."[54]

Other hydropaths were similarly content with only the roughest evaluation of their therapy. They were hazy about the numbers of people restored to health by water cure. None thought of documenting with numbers the claim that a water-cure practitioner could dry up half of the medical practice of a town in a short time, simply by teaching people how to prevent disease. Some invoked the authority of "statistics of all ages," but since few provided any specific data from their own experience, no consensus could be reached on even such elementary matters as how many baths per day to prescribe or how many glasses of water should be drunk.[55]

Nevertheless, some did find uses for numbers in promoting hydropathy. The editors of the *Water-Cure Journal* at various times solicited figures on the numbers of patients, diseases, recoveries, and deaths at the different water-cure establishments. Their purposes were hardly scientific. Rather, the data were intended to herald the approach of the new medical millennium and to "astonish our friends of the Allopathic school . . . The Hydropathic practice is destined not only to surpass, but to swallow up or wash away, every other medical system now existing among men."[56]

Several water-cure establishments, notably the large institutions at Glen Haven, New York, Brattleboro, Vermont, and Northampton, Massachusetts, responded with tables providing such information as the number of patients listed by sex and age, diseases, recoveries, and deaths. Figures were sometimes given on the extent of the bad habits presumed to have caused patients to seek water cure: smoking, drinking coffee, tea, or alcohol, eating between meals, using a drug or patent medicine, and being inadequately treated by practitioners of some other persuasion. Other reports specified occupations of the clientele.[57] Among some 1,500 individuals treated at Glen Haven between 1851 and 1855 were

Housewives	345
Working women	504
Clerks	111
Mechanics	87
Farmers	165
Teachers	94
Ministers	94
Lawyers	99

Given their rejection of orthodox therapies, it is not surprising that hydropaths, like homeopaths, had a low opinion of orthodox statistical efforts. The mortality figures compiled in large cities, they thought, were interesting chiefly because they testified to the inadequacy of allopathic medicine. Otherwise, gathering such information was seen as futile and useless. R. T. Trall, summing up the hydropathic view, saw the allopaths as "excellent statisticians, but miserable philosophers. They can more easily accumulate a mountain of details than establish a single principle." Besides, he concluded, their figures were superfluous: "Why do not physicians, instead of making endless computations of the evils of ignorance, teach us wisdom? Why do they not, instead of piling up mountain heaps of statistics, about the particular manner and way in which lives are lost from ignorance of the laws of life and health, tell us precisely what those laws are? . . . The regular faculty is ever eloquent with the records of mortality, but never ready with words of instruction."[58]

For their part, orthodox physicians found it hard to take

hydropathic statistics seriously. As one editor remarked, it seemed as if their "main object is to catch customers." Beyond that, the regulars concluded that the kind of people who characteristically went to water-cure establishments almost guaranteed favorable statistical results. Unlike city hospitals, the establishments rarely had the most acute cases, the seriously ill. Rather, they attracted the nervous, dissipated, or run-down; individuals with "disorders incident to a sedentary life, or want of attention to the skin, or luxurious habits"; persons who, having failed to observe good hygiene, had looked to patent medicines or to a different therapy of the day to rescue them.[59]

The overall response of regular physicians to water cure was much the same as it was to Thomsonianism, homeopathy, and eclecticism—a mixture of irritation and amusement, anger and incredulity. They marveled that hydropaths seemed so impervious to ridicule and were so persevering in advancing their sect. They considered them naive in advocating a single therapy so exclusively, and obtuse in failing to acknowledge that water was an age-old treatment. As water-cure establishments proliferated around them, orthodox physicians found it difficult to believe that there could be enough people to patronize them all. True, there were many failures and closings of such establishments, particularly during the panic of 1857 and the Civil War years. Nevertheless, speculators continued to regard the institutions as attractive business risks: "Gentlemen of property embark their capital in them, as they would take stock in a railroad, for the sake of the large expected dividends, not caring a fig for the patients beyond their ability to pay the tariff of charges for being watered inside and out."[60]

At the same time, some of the orthodox spokesmen perceived that the hygienic regimen of the water-cure establishments fulfilled a legitimate medical need. Austin Flint, for one, considered it a real misfortune that the regulars had left this general type of enterprise, the "sanitary retreat," to the hydropaths and other irregulars. Still, as in the case of homeopathy, the exclusivity and arrogant manner of water-cure enthusiasts gave the sect a bad odor which effectively obscured for many its potential medical value. The competition from hydropaths undoubtedly led conventional physicians to prescribe more visits to mineral springs or health resorts for their own patients. However, the regular medical profession as a whole was far too insecure to admit publicly that anything connected with irregular medicine was any good.[61]

7

Statistics of Mind and Madness

One matter that regular physiologists could share with sectarians and other health reformers was the conviction that improving a person's physical health was only one side of the goal of total hygienic well-being. Just as important was the promotion and maintenance of mental health. As the Hartford physician Amariah Brigham summarized it: "The whole man should be improved. Not only should all his powers be developed, but they should be developed harmoniously."[1]

This aspect of health reform manifested itself in a widespread enthusiasm for a number of new techniques which promised scientific illumination of the mind and brain. It also found expression in a concern for the presence of mental disease in society, in the provision of institutional means for coping with that ailment, and in efforts to determine its causes. The pursuit of these various concerns and interests, combining as it did the approaches of political arithmetic, Louis's numerical method, and empirical social science, represented a major phase in the extension and maturation of midnineteenth-century medical statistics.[2]

Mesmerism and Phrenology

Two techniques for a time held promise of providing insight into the behavior of nineteenth-century Americans. One of these innovations, animal magnetism, or mesmerism, rose and fell so rapidly in the estimation of the medical profession that it was hardly even mentioned in contemporary medical literature. Certainly there were no antebellum attempts even to measure the extent of mesmerism-hypnosis—and only a few studies of its possible medical uses. The lectures and experiments of the French layman Charles Poyen and his followers in New England, together with the writings of the Englishman John Elliotson and various Europeans, stirred up some medical interest in mesmerism during the mid-1830s, but it did not last long. The popularization of mesmerism by such impressarios as Rubens Peale, in whose museum public demonstrations of it were given, and the association of some mesmerists with Swedenborgianism led to an increasing skepticism of the practice.

By the midforties the measure was being left mostly to quacks. Daniel Drake encountered nearly as many mesmerists in the Mississippi Valley during this time as he did Thomsonians or other botanical practitioners, and most of them he judged as mountebanks. "Things are coming to that kind of pass," he reported, "that we shall soon not be able to distinguish the pass of an imposter from the pass of a scientific Mesmerizer." By 1845 in Cincinnati even the quacks had given up the practice, which was all the more unusual, Drake thought, because more than almost any other therapy, mesmerism encouraged "tender relations . . . it requires the doctor and the patient to be of opposite sexes." Josiah Nott, a regular physician, regretted that mesmerism had been so discredited by quacks that it had never been properly tried out by the regular medical profession. Nott himself experimented on some fifty subjects over a three-year period in Mobile, Alabama. His observations left him with "no ground whatever for reasonable doubt" of the reality of the hypnotic process and of its possible medical usefulness.[3]

Almost the only other American to make any kind of systematic tests of the effects or claims of mesmerism before the Civil War was the Philadelphia physician John Kearsley Mitchell. Mitchell, who hoped to explain mesmerism both mechanically and psychologically, experimented together with a friend on about 150 subjects over a five-year period. Some of these tests were conducted on private patients, some on inmates at the local orphanages and asylums for deaf-mutes. The report of these studies included tables summarizing the effects of mesmerism on the pulse, respiration, and body temperature, and another table on time factors involved. Other tests concerned the extent of the individual's suggestibility to the process. In a trial on 36 children, Mitchell found that 25 were not affected at all, while 11 went into different degrees of hypnotic sleep, though only 5 of those were substantially affected.[4]

Another European import, and one which most Jacksonians agreed was more promising than mesmerism, was phrenology, the study of mental faculties through the shape and size of the skull. The course of phrenology in midnineteenth-century American medicine roughly followed that of mesmerism, but its vogue was far greater. In fact, in the two decades before 1840 it was popular both among orthodox physicians and irregulars. Though the regulars began backing away from the practice as laymen and quacks took it up, the overall statistics on its early growth were impressive.

The Boston physician J. V. C. Smith pulled together many of these data in 1840 for his *American Medical Almanac*. Although no count of practitioners or followers of phrenology was available at that time, it was generally agreed that the practice had a large following that included professionals as well as laymen. The huge sales of books on the subject gave some indication of the number of believers. The basic works of Franz Joseph Gall and Johann Caspar Spurzheim, Smith found, had "a constant sale," the latter especially in their American editions. Various editions of George Combe's *Constitution of Man* were estimated to have a sale of twenty thousand copies in the United States, and Alexander Combe's phrenological *Principles of Physiology* sold over thirty thousand. American publications proliferated after the appearance of Charles Caldwell's *Elements of Phrenology* in 1824 and most sold steadily. *Fowler's Practical Phrenology*, which Smith regarded as the best elementary or popular work then available, had sold ten thousand copies in five editions by 1840, and Fowler's phrenological almanac was selling at least that many annually. At least two native phrenological periodicals were being published, and many of the standard American medical texts were paying some attention to phrenological concepts.

The demand for lectures and demonstrations on phrenology was considerable even before Spurzheim's triumphant

American visit in 1832. By the time of George Combe's visit later in that decade, the thirst had become insatiable. In a small town near the frontier, Charles Dickens saw people flocking to the phrenological lectures of a "Doctor Crocus."[5] In Boston the speakers on phrenology were more often respectable physicians, including Samuel Gridley Howe, John C. Warren, and Elisha Bartlett. In New York, the affluent were visiting Orson and Lorenzo Fowler's offices for personal character analyses and consultations. Meanwhile, the handful of phrenological societies which were organized in the early twenties had multiplied to several dozen by 1840, though these groups were notoriously transitory.

The Fowler brothers, energetic publicists and showmen who turned phrenology into a household word as well as into a profitable business, perceived very early that the United States, with its great variety of races and nationalities, offered an unusually good milieu for the study of phrenology. Besides, Americans were infinitely curious about themselves, gullible, and had a great desire to be entertained. For the individual who could afford it, the measurement and phrenological assessment of the skull provided one with an intriguing and usually flattering psychological profile. For the public as a whole, published analyses of the heads of writers, politicians, and other public figures were good sources of conversation and gossip. For the scientist, phrenology seemed, for a time anyway, as worthy of study as chemistry, botany, or any other field. To be sure, American scientists of this period did not add much to the original concepts of "cerebral localization" set forth by Gall and Spurzheim, but some of them refined the original classifications and nomenclatures, combined phrenology with their anthropological studies, or attempted to correlate its principles with the emerging social sciences.

The number of mental faculties to be measured by phrenology varied from one writer to another. Gall originally differentiated twenty-seven, each of which, he thought, could be measured in a given individual by reference to the relative prominence of a corresponding segment of the skull. For the most part, the same twenty-seven faculties were retained by Gall's successors, but several introduced new faculties and various changes in nomenclature. Spurzheim and George Combe both evolved systems with thirty-five faculties, for example, and in the United States, Caldwell proposed thirty-four. The Fowlers at first perceived thirty-seven faculties in 1836, but this number grew to forty-three, and they made a number of changes in the nomenclature. Joseph R. Buchanan felt that additional subdivisions of the brain and much greater numbers of faculties were necessary "to constitute a portrait of human nature."[6]

In classifying the mental faculties, the Fowlers tried to approach the original perfection of God's classification, which, of course, they could only dimly perceive. Their solution was simply to "let the faculties *classify themselves* according to the grouping together of their respective organs in the head." The Fowlers' middle-class customers were not that interested in knowing just how the traits were classified. The value of phrenology to them was that it provided a means to make novel, even daring, pronouncements about the individual psyche. Presumably it could reveal aspects of a person's character that previous generations had diligently striven to keep private. Phrenology was popular, therefore, because its revelations were often highly titillating.

Its nomenclature differed little from the vocabulary of popular novels. Attuned to middle-class values and morality, it ran the gamut of human sentiments, base and noble. Jacksonians thus eagerly sought to verify their personal moral development by inspecting such faculties as benevolence and spirituality. They were equally eager to learn about each others' propensities for "amativeness" and "philoprogenitiveness." They also avidly measured the relative amounts of self-esteem, acquisitiveness, or destructiveness revealed in the skulls of their worldly leaders. Phrenologists believed strongly in human perfectibility and maintained that with proper exercise any mental faculty could be enlarged. The ideal, of course, was a "good intellectual head," in which all of the faculties were balanced.

Not at all surprisingly, phrenologists considered "calculation" among the essential qualities which all Americans should develop. The Fowlers especially urged that people "embrace every opportunity"—while walking, riding, or at work—to cultivate the capacity for performing mental arithmetic. As one exercise, they urged travelers speeding along on a train to figure out what any given velocity would mean in terms of miles per day or year. Peddlers or merchants were encouraged to learn to make their business transactions in their heads. Pencil calculations were to be done only as a check. "You should also *charge* your memory with numbers," the Fowlers advised. "You would then seldom be at a loss for statistical information—the most difficult matter to be recollected."[7]

As a central aspect of their extensive phrenological activities, the Fowlers developed the technique of taking casts and making busts of living heads. By 1840 they had over two hundred busts of leading or notorious Americans, which they used to demonstrate the external manifestations of the various mental faculties. They had also collected numerous skulls, both animal and human. Other individuals and institutions around the country built up similar collections. Those of the Boston, Albany, and Buffalo phrenological societies were especially large. The Yale Medical School displayed George Combe's specimens in a permanent exhibit, following his American tour, and the Philadelphia Academy of Natural Sciences acquired a substantial part of the largest American collection of skulls, that of Samuel George Morton.[8]

Phrenology, Craniology, and Race

The skull measurements that such collections facilitated and that dominated the practical or popular phrenology of the Fowlers varied in importance for American physicians. Relatively few seem to have rejected phrenology outright during this early period, though many were interested in using it only to investigate mental processes. Other scientist-physicians, however, became predominantly concerned with its bearing on craniology. For them, skull collections provided not only bases for analysis of the human brain but also quantitative resources for broad comparative studies of human races in the tradition of such European investigators as Johann Friedrich Blumenbach or James Cowles Prichard.[9]

Since colonial times, white Americans had been examining

the similarities and differences between themselves, blacks, and Indians. Some, like William Byrd, John Mitchell, and William Charles Wells, looked for meaning in differences of skin color. Others, like Jefferson and Albert Gallatin, looked at linguistic forms. And, with the opening up of burial mounds in the Ohio Valley, with the expanding contacts of American explorers and travelers with other cultures around the world, and with the establishment of phrenological collections, comparing skulls became a significant element in the overall anthropological discussion. Such prominent physicians as Charles Caldwell and Josiah Nott became intimately involved in philosophical aspects of the argument, notably the question of the original unity of the races, and Samuel G. Morton of Philadelphia undertook the careful quantitative studies of skulls that provided a base for much of the subsequent discussion.

Morton began collecting skulls during the 1820s. By the thirties he had enlisted a broad segment of the American scientific community in helping to add to his collection, which eventually included over a thousand human crania and many from animals. Among the most helpful contributors were the Indian specialist Henry R. Schoolcraft in Michigan, the scientist William Maclure in Mexico, and the United States consul George Gliddon in Egypt, along with missionaries and explorers in a variety of places. Doctors such as Samuel P. Hildreth and Benjamin Tappan sent crania from mounds and caves in the Ohio basin. From army posts along the frontier the medical officers Zina Pitcher and Eugene Abadie furnished skulls of recently deceased Indian warriors. Physicians who went out to the colony of Liberia shipped back skulls of native Africans, while naval surgeons—among them William Ruschenberger, Henry S. Rennolds, and Waters Smith—obtained crania from Peruvian cemeteries.

Morton's first skulls were used to illustrate his lectures on anatomy in Philadelphia. As his collection expanded he became increasingly interested in comparative studies of the different races, not only according to the exterior form of the skulls but to their "internal capacity, as indicative of the size of the brain." In addition, however, he was interested in studying the pathological indications of disease or injury and, to a lesser extent, their phrenological significance. Morton numbered each of his skulls; he measured such things as the diameter, circumference, and facial angles and determined the internal capacity in cubic inches by means of a device invented by his friend and colleague John S. Phillips; he also recorded information about the date, the place in which it was found, and other details.[10]

Morton then classified the skulls, following Blumenbach, within five large racial groups: Caucasian (subdivided into 7 families), Mongolian (5), Malay (2), native American (2), and Ethiopian (2). Classification, he thought, could no longer pretend to reflect some divine arrangement. Rather, like the new generation of nosologists, he considered such arrangements necessary for the orderly study of anthropology. His was a mixed classification based both on the linguistic traits of the various races and on the capacity and conformation of the crania.

Although Morton later refined his views in works based on larger numbers of skulls, his fundamental conclusions were evident by 1839, when his most influential publication, the *Crania Americana*, appeared. There, in a careful setting of cultural and historical detail, Morton presented an exhaustive analysis of 147 skulls representing forty different Indian tribes or nations from both North and South America. From their study he concluded that the "American Race" differed from all of the other races, and he found nothing to sustain the claim of a single original race. To support his views he combined the data in tabular form, particularly the anatomical and phrenological measurements of the skulls. In an appendix to the book, George Combe provided a phrenological interpretation of the data. Morton thought that phrenology was still unproved and not yet suitable for use in his scientific conclusions. However, he felt that his work did contribute to the scientific evaluation of that method. It was prudent, he thought, "to present the facts unbiased by theory, and let the reader draw his own conclusions . . . Yet I am free to acknowledge that there is a singular harmony between the mental character of the Indian and his cranial development as explained by Phrenology."[11]

Combe's appendix on phrenology almost certainly tended to discredit the whole book in the eyes of some readers. At the same time, Morton's views on the distinct origins of the races ultimately made the book suspect in another way, for they placed it squarely at the center of the growing political controversy over the role and position of the Negro. Nevertheless, many of Morton's contemporaries were impressed by his "monumental data," and some even saw him as "the undisputed founder of a *new science* [comparative anatomy], which was first suggested by Blumenbach."[12]

Physicians and laymen alike continued to show interest in the measurement of skulls during the middle decades of the century. When Daniel Webster died, it seemed only natural for medical and popular periodicals alike to publish details of the statesman's cranial cavity and brain size. Webster admirers were probably not much surprised to learn that the circumference of the great man's head was 23³/₄ inches; Napoleon's had been 23. Americans probably took much satisfaction in knowing that Webster's brain ranked among the largest ever recorded, with only Cuvier's being heavier.[13]

The Measurement of Society

The analysis of the skull of a single prominent individual, though interesting, was of minor importance beside the quantitative analysis of many skulls. If some, like Morton, focused their attention on the physical measurements of skulls, others in antebellum America were more interested in measuring the moral aspects. A few even began to perceive that the study of such collective data about humans could lead to a science of society.

The Boston reformer Samuel Gridley Howe was one who found both scientific and social values in phrenology. For him, being able to compare a given brain with an established standard was a great advantage in determining idiocy or other defects. However, the moral significance of the large accumulations of skulls seems to have left the most powerful impression on him:

> I know of nothing . . . more startling, than the array of skulls and casts, in the great Phrenological Golgothas of London and Paris:

first, you see hundreds of mean, flat skulls, with the top as narrow, and the crowns as low, as those of monkeys or dogs; they are the skulls of idiots, they are ranged side by side, in long rows . . . then there are others, with the back parts of the head very large, but the front, villanously low—these were gluttons, or sensualists, robbers, pirates, or murderers; they are ranged rank upon rank, by scores, and by hundreds; others have the head better proportioned, but the upper region is narrow or flat, and the part above the eye-brows projecting, and well developed; these are the knowing ones, the shrewd, cunning, cautious, cheats or swindlers—of these are multitudes; other ranks, more thinly filled, show you the lofty top, the bulging front, the spacious breadth, which mark the heads of philosophers, the moralists, the men who are a law unto themselves.[14]

Howe's fellow New Englander Elisha Bartlett shared some of the preconceptions which led to these broad generalizations. But he carried Howe's impressions a step or two further. Bartlett welcomed the emergence of this "science of the mind," though the process of collecting facts about its broader uses and values was still continuing in the late 1830s. He agreed with the idea that phrenology provided a broad framework for the making of moral judgments, one that might bring a greater measure of objectivity to that function, but in addition, he pointed out some of its potential social applications. Like Horace Mann, Bartlett thought that the study of phrenology could shed light on the processes of learning and education. And, in many other ways, he felt that the science could be valuable if its findings about individuals could be extrapolated to society itself. Bartlett was impressed with the phrenological doctrine that every mental faculty could be strengthened or manipulated, and he applauded the efforts being made by society to provide inmates of mental institutions with morally uplifting environments. Even more, he was convinced that the extension of phrenological principles from the single person to the infinitely many, "from the individual to the race," would open up endless possibilities for generalization about human behavior. "Beautifully unfolding itself in the process of this interpretation, shall we find, every where, *Law*. Chance disappears, and we see that, throughout all that multitudinous thought and action of humanity, constituting its history . . . is there nothing fortuitous, nothing accidental, nothing anomalous."[15] Laid forth in such a systematic and orderly way, the study of society would become truly scientific.

Bartlett was by no means alone at that particular time in envisioning an authoritative role for phrenology in the analysis of society. However, it ultimately became clear that the particular nature of phrenology did not lend itself to such a role. Far more appropriate to the study of social problems was the tool of statistical analysis. In every advanced country, reformers were already busily collecting empirical data, not only of disease, but of crime, bastardy, divorce, pauperism, education, and other social phenomena. In France, in the midst of such ferment, August Comte provided a basic rhetoric and terminology for sociology, and he outlined its structure, but it was the Belgian astronomer and demographer L. A. J. Quetelet who did most to turn social theory into a true social science.

Quetelet did not, perhaps, accomplish much more than some of his contemporaries in building up a network for collecting the vital, anthropological, and social data that interested him, but he did go beyond them in the statistical knowledge and methodology which he brought to bear upon the amassed information. Quetelet showed how higher mathematics could become an integral part of applied statistics. Largely through his demonstrations, for example, the bell-shaped graph of the normal distribution of variations from the median eventually became almost as familiar to those working for social, political, and medical change as it had been to mathematicians. In such works as his *Sur l'homme* of 1836, Quetelet updated concepts of statistical order in human affairs, such as William Petty's political arithmetic and William Derham's physico-theology, for the nineteenth century. His was a distinctly realistic view, and one that provided a needed corrective to the romantic individualism of the age.[16]

Few Americans before 1850 had the mathematical competence to conduct their statistical studies with the sophistication of Quetelet. Nevertheless, many, particularly physicians, became familiar with his techniques and findings. They quoted him frequently and applauded the new scientific element which his quantitative conclusions brought to studies of man and his moral, physical, and medical condition. By the 1830s, American readers were reading Quetelet's early works—on the Belgian census, on suicide and crime, on man's height and growth. By the forties he was well known here as an international leader in science, and Americans had joined with scientists elsewhere to supply data for his studies. Some sent him meteorological and oceanographical data. Others exchanged reports of vital statistics with him and furnished anthropological information. Not surprisingly, when George Catlin visited Brussels in 1846 with a collection of his paintings and a dozen Ojibway Indians, Quetelet seized the opportunity to measure the Indians.[17]

American reformers and physicians gradually became familiar with Quetelet's interpretations of vital data in terms of the statistical mean. So far as human studies were concerned, this involved becoming acquainted with the concept of the "average man." For Quetelet, the average man was to some extent an ideal. More often it was a composite being whose traits and patterns of actions were hypothesized from many means calculated from real men's lives. In the dynamic environment of Jacksonian America, of course, the concept of the average man had none of the broad currency of the notion of the "common man." Budding social scientists could only speculate about a day when they would have data that was complete and extensive enough to portray an average American. For the time being, most of their concern centered on learning about and dealing with the atypical members of American society: the deprived, the deformed, the criminal, the vanishing races, and so on. Of these unfortunates, none received more attention than the mentally ill and all those whose mental well-being seemed to be threatened by conditions around them.

Jacksonian Excitements and Mental Health

The prevention of mental disease was a crucial objective of the health reformers and hygienists. Progress toward this goal was thought possible if one could ensure an external environment for the individual that was without extreme tensions or excitements, and if one could achieve an internal emotional outlook that was well balanced and stable. However, the realities of early nineteenth-century life presented formidable barriers to achieving this ideal.

The constantly shifting population by itself kept American society at a high level of tension. Some people may well have become unbalanced or at least intoxicated by the optimistic spirit and high hopes of the age, but large numbers experienced predominantly negative feelings, worries, and disappointments in making their moves. Anxieties of all sorts haunted those who left farm and village to move to the city. The homesick immigrants in their miserable huts or tenements were frequently noticeably depressed and an equally dark despondency often settled upon Easterners who had uprooted themselves from familiar surroundings in exchange for the multiple uncertainties of the strange, difficult, and often hostile environment of the West. During the 1820s the itinerant clergyman Timothy Flint frequently encountered intense gloom and regret among New Englanders who had recently moved into the Ohio and Mississippi valleys. As he thought about the problem, he wished that it were possible to devise a statistical measure of the happiness of such individuals before and after their moves. Such a calculation, he suggested, would be invaluable in helping others to make up their minds about emigration.[18]

Actually, things were sometimes just as unsettling for those who stayed at home. In the 1830s the hectic pace of Jacksonian life was both marveled at and complained about, and as cities grew, the numbers of comments about it accelerated. There were just too many stimuli and too much excitement for all but the toughest individuals, particularly in the cities. Urban editors and physicians pointed to the detrimental effects on the nervous system of the increasing propensity for "fast walking, fast driving, fast eating and drinking, fast bargains, fast business . . . fast everything but *fast-ing!*" In New York, the preacher Henry Ward Beecher found that the "bustle of the street, the ceaseless thunder of the vehicles, the rush to-and-fro of multitudes of people" was more than many of his congregation could bear. If America were an autocracy, he mused, he would be in favor of "having a registration of the people in the city, and having all those who are nervous turned out into the country."[19]

One of the most conspicuous and deleterious facets of this fast-paced life was the constant overriding pressure of the work ethic. For some, the insatiable desire to accumulate property seemed to be the principal "characteristic of the age." They saw this "money-making mania" as an obsession which allowed no time for relaxation and which drove the infected until they dropped dead or collapsed of nervous and physical exhaustion. Administrators of mental hospitals agreed that this obsession was the underlying cause of insanity in many of their patients, while business failures or losses showed up prominently among the reasons reported for suicides.[20] Beecher was shocked by the enormous figures of business failures in New York during the 1840s and 1850s and by the numbers of men who were destroyed mentally or physically in the process. He was especially appalled by the proportion of his acquaintances whom, over a thirteen-year period, he had seen "broken down, driven into the lunatic asylum, driven into the hospital, or driven into the grave, by the mere effect of exhaustion, of overtaxation, of incessant labor." Even among successful businessmen, he conservatively estimated that over 50 percent eventually succumbed to idiocy, insanity, "softening of the brain," or "nervous exhaustion."[21]

Just as upsetting for many others was the rush of words and ideas which bombarded them from many directions and in rapidly accelerating quantities. Amariah Brigham, Superintendent of the Hartford Retreat, wondered at the extensive agencies which had already come into being before 1832 to disseminate ideas. It was, he thought, "fearful to contemplate" the high pitch of artificial excitement which was being maintained in people of his generation by the constant impact of newspapers, novels, lectures, sermons, and tracts of one sort or another. In Hartford alone, the minds of the seven thousand citizens were kept stirred up by seven local political newspapers and five religious weeklies, by the many schools and Sunday schools; by books in family libraries; by the weekly lectures and debates at two lyceums; and finally, by the two or three Sunday services conducted at each of the ten local churches plus the twenty or thirty midweek religious meetings held in the city.[22]

As he developed his ideas during the early thirties, Brigham came to regard religious meetings as having the most noxious effects on health of any of these agencies. He focused on religious activity partly because most of the midweek prayer meetings, as well as those organized for the benefit of colonization, Bible, missionary, abolition, tract, or temperance societies, were held in the evening; frequent exposure to the night air and to drafty church parlors was regarded as a certain invitation to tuberculosis and other maladies. Where night services had been virtually unknown about 1800, by 1835 the church parlors were being filled more than half the evenings of the year. And the trend seemed to be particularly dangerous for females, Brigham noted, not necessarily because they were more delicate, but because they generally composed more than half the audiences.

Added to the problem of exposure to night air was the even graver matter of the high level of mental excitement stirred up by the constant exhortation of preachers at such meetings. This was, of course, especially apparent at revivals or camp meetings, but Brigham asserted that even at regular services congregations were habitually kept tense and anxious because the majority of American preachers practiced the "denouncing" or hell-fire type of sermon. Given the immense number of religious meetings, the cumulative effect was a dangerous amount of excitement that often led to insanity, again more frequently in females. Striking against the very foundations of the enthusiasm for improvement of the times, Brigham used a rough quantitative argument to urge a gentler form of religion in the interests of health. The plethora of evangelistic evening services, he thought, encouraged "a kind of theatre-going spirit, i.e. a love of excitement, incompatible with a love of domestic life." In fact, to the consternation of the religious establishment, he made the astounding assertion that going to such services was even worse for people than such traditionally reprehensible activities as going to plays or dancing:

I consider theatres and balls as less injurious to the health of the people of this country than religious night meetings, because attendance on the former is far less frequent. Probably not more than one female out of two or three hundred attends a ball in a year, or at the most not more than three or four, and of the same number not more than four or five go to the theatre more than half a dozen times in a season, and these are confined to the large towns. But throughout the whole community, at least 25, if not 50 out of one hundred females, between the ages of 15 and 50, attend religious meetings at least 100 or 150 nights a year.[23]

Brigham's conclusions naturally stirred up controversy, especially among his New England neighbors. But he was by no means alone in remarking on the excesses of revivalist preaching and the mental excitement which they produced. In fact, throughout the Jacksonian era the statistical reports of mental hospitals consistently identified the excitements of religion, along with those of a variety of social phenomena, as prominent causes of insanity. The Hartford Retreat, for example, attributed insanity to such causes as "preaching 16 days and nites," "Fourierism," "anticipation of wealth," "Mormonism," "study of Phrenology," and "anti-rent excitement." During the early forties mental hospitals admitted dozens of individuals whose derangement was supposedly caused by the end-of-the-world doctrines of William Miller, and bills of mortality reported large numbers of suicides from the same cause. Brigham felt that Millerism, spread by the printed word, "would have done no harm, if there had been no *preaching* of the doctrine,— *no nightly meetings and collecting in crowds to hear and see* . . . According to our observation, the greatest number [of those made insane by Millerism] occur among those who have long been pious, but who having become excited, agitated, and worn down by attendance, week after week, on nightly religious meetings, until their health became impaired, they then began to doubt their own salvation, and finally despaired of it, and [became] decidedly deranged."[24]

Brigham's medical argument for a more subdued and humane regimen of church-going was paralleled in the thirties and forties by a religious argument. Brigham's Hartford contemporary and neighbor, the preacher Horace Bushnell, particularly urged that conversion should be accomplished through a steady, organic process of Christian nurturing in the family rather than by a sudden spasm of feeling in the charged atmosphere of a revival. Health reformers, for their part, fostered a similar concept of the family unit as the organic focus of education, religious belief, and moral attitudes, as well as of enlightenment on matters of hygiene.

The editors of the *Journal of Health* offered a number of practical suggestions for helping individuals cope with mental excitements. One solution proposed that every parent should try to control the problem within the family. As an aid to achieving this control, they suggested keeping track of the traits of individual family members, not by phrenological readings, but through a "moral thermometer," possibly suggested by Rush's scale of temperance and intemperance (see Figure 2).

The editors recommended that the scale be hung in a conspicuous place and that family members keep daily records of their emotional reactions, which were then to be related to meteorological readings. "They will thus be enabled at any time by looking back to former entries, to see what measure of improvement has been gained in a given period, as well as to discover, by comparing the entries in the moral with those in the meteorological journal . . . whether that which is said to affect so powerfully the human temper, the state of the atmosphere, does indeed cause all the alterations we remark."[25]

However salutary such preventive measures may have been, there was no escaping the fact that most emotionally upsetting conditions were well beyond the capabilities of families, physicians, or moralists to remove or control. The question of what to do with those who had actually become mentally disordered as a result of such stress and excitement, or from some other cause, thus remained. Provision for these unfortunates became a major social, medical, and statistical concern.

Insanity in the Age of the Asylum

At the beginning of the nineteenth century, provisions for the mentally ill were varied and often haphazard. A large number of such unfortunates were cared for at home. Many roamed

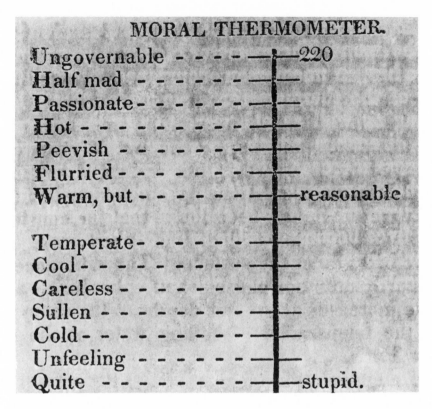

MORAL THERMOMETER.
Ungovernable - - - - ——|——220
Half mad - - - - - - - —|—
Passionate - - - - - - - —|—
Hot - - - - - - - - - - - —|—
Peevish - - - - - - - - —|—
Flurried - - - - - - - - —|—
Warm, but - - - - - - - —|——reasonable

Temperate - - - - - - - —|—
Cool - - - - - - - - - - —|—
Careless - - - - - - - - —|—
Sullen - - - - - - - - - —|—
Cold - - - - - - - - - - —|—
Unfeeling - - - - - - - —|—
Quite - - - - - - - - - —|——stupid.

Figure 2. A "Moral Thermometer," from the *Journal of Health*, 4 (1833), p. 5.

the countryside, while still others were incarcerated in alms-houses or jails. Apart from Benjamin Rush, few physicians gave any special attention to the insane in 1800, and there were not half a dozen institutions providing medical care for this ailment.[26]

In the three decades after 1800, a few additional hospitals for the mentally ill were built, notably the Bloomingdale Asylum in New York City, the Friends Asylum at Frankfort, Pennsylvania, the McLean Hospital near Boston, and the Connecticut Asylum in Hartford, eventually called "The Retreat." The early institutions grew out of local needs, the extent of which was sometimes established, as in Connecticut, by rough statistics drawn from questionnaires. By the late twenties the experiences and statistical reports of these institutions, along with those of foreign asylums like the Retreat at York, England, were stimulating the demand for such facilities in other localities. In New York, T. Romeyn Beck called for a statewide system of publicly supported asylums for the insane, all provided with proper medical staff, in contrast to the inadequate provisions of almshouses. There, he thought, "instead of merely . . . keeping [the insane] in custody, to wear out a miserable existence, new trophies might be gained for the medical art, and many valuable citizens restored to their families and the community."[27]

In 1829 the Massachusetts legislature authorized Horace Mann to survey that state's insane. Mann's report, which drew attention to the extent of insanity elsewhere, notably England and Connecticut, as well as in Massachusetts, proved to be a decisive influence the next year in obtaining authorization to build a state mental hospital. When the hospital was completed in 1833, the highly statistical reports of its first superintendent, Samuel B. Woodward, stimulated even more interest in the insane among reformers in Massachusetts and in other states.

In 1835 reformers estimated that, nationwide, one person in every thousand was insane, which meant that there were approximately twelve thousand mentally ill individuals. Of these, hardly a thousand were then in asylums, and perhaps only another three thousand were reasonably well cared for at home. The editor of the *Boston Medical and Surgical Journal* was disturbed by the figures for those who remained: "It is a melancholy reflection, that between seven and eight thousand of these unhappy, unconscious, irresponsible fellow beings, are paupers—many of whom roam over the country, neglected and often abused, without exciting that active benevolence of feeling in the whole community which is imperiously called for by this great multitude of wretched lunatics, who should all be housed, be fed, be clothed, and treated in the kindest manner at the expense of the States in which they have had a habitancy."[28]

The growing humanitarian concern for the mentally disturbed generated considerable demand for more accurate information. By 1840 the demand bore fruit in the federal census which provided for the first national enumeration of the insane and the feeble-minded. When completed, the enumeration, despite many errors, was a major stimulus to local reformers and officials to look more closely at the problems and at the statistics.[29] In fact, by 1840 the effort to determine the numbers of these unfortunates, to get them out of the prisons and garrets, to build humane institutions for them, and to restore as many as possible to sanity, had already built up considerable momentum.

In 1841, Pliny Earle showed that the United States had indeed been far from backward in making provision for the insane. Only France, with thirty-four asylums, seemed to have done more for them. England, with sixteen asylums, was then third. Of the nineteen asylums that he counted in the United States, ten were state institutions, two were municipal, and seven were private. Earle and other American observers were persuaded that many of those institutions already in operation "favorably compare with the best in Europe." However, they recognized that the existing asylums did not begin to meet the local needs, and they deplored the slowness of some states in making any provision for such structures. In 1845, eleven of the twenty-six states still had no public or private insane asylums. However, within six years, virtually all of them had remedied this lack, thanks to a new wave of crusading fervor stimulated by Dorothea Lynde Dix.[30]

The earlier efforts to provide for the mentally ill were essentially local, with the impetus coming from local reformers, legislators, and physicians. Dorothea Dix, a frail New Englander, made the cause a national one during the forties. Probably it was the spectacle of a well-born lady venturing to poke around in jails and poorhouses, as much as anything else, which first brought her the public's attention in Massachusetts. But her persistence, her unswerving sense of indignation, and her stark statistics were what influenced legislators from Rhode Island to Illinois, from New York to Alabama. Dix's reports, or "memorials" to these legislators, included the details of her town-by-town surveys of the respective states backed up by published data from existing asylums. In each case, she pointed out, these "statistics laboriously collected exhibit plainly your need." Much like John Howard two or three generations earlier in England, she used the solid weight of numerical findings to buttress her state-by-state appeal to the sentiments of "justice and humanity."[31]

Once established, most antebellum asylums were administered by physicians who had had no special training in psychiatry. However, almost to a man, they adopted the hopeful moral treatment of Philippe Pinel and of Daniel and William Tuke for their institutions in place of the older repressive concepts. Moral therapy—the improvement of the patient through scrupulous hygiene, a pleasant environment, and humane treatment—was a laborious and complex matter, but Americans took to it enthusiastically. To carry out its precepts, the asylum superintendents became organizers par excellence, systematizing the asylums' procedures to a high degree, and committing themselves to the use of statistical analysis. In offices that were veritable registries they supervised careful record-keeping not only of business details but of facts about the patients and their maladies. Not a few of them took the time to keep meteorological observations and to observe to a greater or lesser extent the influence of the seasons on the various mental conditions.

Asylums were expected to keep basic information on admissions, discharges, age, sex, periods of illness, and mortality of the patients, along with various clinical information. In all likelihood, they did a better job of this than the early American general hospitals.[32] At any rate, the Association of Medical Superintendents of American Institutions for the Insane, upon its formation in 1844, made a conscious effort to promote the gathering of such data in the various asylums by creating a standing committee on statistics.

Many superintendents faithfully tabulated matters such as the occupational backgrounds of patients, even though some recognized that such figures did not provide an accurate basis for judging mental disease in the whole population. Worcester Hospital records for the years 1837 to 1841, for example, showed that for male patients the five largest job categories were consistently farmers, laborers, shoemakers, seamen, and merchants, in that order. At the Bloomingdale Asylum between 1821 and 1844, and at the Pennsylvania Hospital for the Insane in 1845, most male patients were farmers, merchants, clerks, laborers, and carpenters.

Among the most important data, of course, were statistics pertaining to the causes of insanity in a community. Asylum superintendents proved to be among the most enthusiastic classifiers of disease. In common with medical practice everywhere, their reports in the early decades of the century tended to characterize causes as "remote" or "proximate," "predisposing" and "exciting." By the 1840s, these vague classifications were being dropped, and instead some institutions were separating cases according to the familial or environmental origins of the patients. Most commonly, however, patient populations were divided into those whose illnesses arose out of physical causes, by then the larger group, and those from moral causes. Within these two broad categories, most of the hospital cases were attributed to five more or less specific causes. Out of a total of 1,196 patients admitted to the Worcester Hospital between 1833 and 1840, over three-fourths were put into these five groups, as follows:[33]

Domestic afflictions (love, family deaths, financial)	294
Chronic ill health	243
Intemperance	190
Masturbation	107
Religious influences	85

Of these various causes, Worcester officials felt that some—ill health, domestic afflictions, and religious influences, as well as hereditary conditions—were largely beyond the current power of society to prevent or control; hence, there would be a continuing need for asylums. However, certain leading causes, notably masturbation and intemperance, seemed to be "susceptible of *immediate* and final suppression."[34]

The publication of regular annual reports presenting such data in detail was a matter of great importance to the asylums, just as it was to other medical and philanthropic institutions.[35] The reports served as the primary scientific media communicating the asylum's medical experiences, and administrators went to great lengths to exchange their publications with those prepared by their counterparts in this country and abroad. To an even greater extent, however, the reports served the crucial purpose of justifying the institution's very existence in the minds of local legislators, trustees, donors, and the public. As a result, the scientific function of the reports was all too often subordinated or sacrificed to the political; statistical material came to serve publicity more than disinterested inquiry. The eventual exposure of this practice launched a critique of statistical methodology that hit at the roots of antebellum optimism underlying the institutional care of the insane.

Criticizing the Asylum Statistics

Those in the community who noted or used the statistics produced by asylums were interested above all in the numbers of cures obtained by the respective institutions, and superintendents understandably furnished data which would make the best possible impressions. As a result, and because most superintendents had an imperfect understanding of statistical methods, many of the early published data on cures bore little resemblance to reality. One frequent practice was to compute the percentages of cures from the number of patients discharged from an institution rather than from the number admitted; another was to include the number of deaths at the asylums among the figures for recoveries. Particularly when applied to small numbers of cases, such computations readily produced highly impressive curability rates, figures which showed some institutions curing as many as 70–90 percent of their patients.

Another factor that ensured an apparently high rate of curability was the careful selection of patients. Asylums that accepted only recently identified cases of illness obviously had better success than those required to take in many hopelessly chronic cases as well. Most institutions had at least some of both categories, though it was not always easy to tell this from the statistics. The Hartford Retreat, in its first ten years, admitted a total of 516 patients. About half, or 263, were old cases, of which recoveries were claimed for 62, or 27.3 percent. Out of 253 new cases, however, 230 recoveries were reported, or 90.9 percent. The Worcester Hospital claimed that out of

1,196 patients admitted between 1833 and 1840, 506, or about 42 percent, recovered. The ratio of old to new cases was similar to that in Hartford, 19.3 percent to 87.3 percent. However, these figures were meaningless wherever uncertainty existed as to how long the illnesses had existed before diagnoses were made.[36]

Overall, the statistics from mental hospitals during the 1820s and 1830s confirmed the faith people had in moral treatment and gave supporters the optimistic news about cures that they wanted to hear. Not all American asylums were as good as those of Europe, but in 1841 nine out of eighteen had curability rates that ranked with the best of the foreign institutions.[37] The value of the statistics in fostering good public relations was confirmed many times over by American communities' willingness to build more and more institutions for the insane. However, superintendents began to perceive that the statistical enthusiasms they had displayed as administrators had outdistanced their statistical skills. During the 1840s several of them suggested much-needed improvements.

Thomas Kirkbride of the Pennsylvania Hospital for the Insane recognized that care had to be taken in making the original diagnostic observations of cases if hospital statistics were to be accurate and worthwhile. Among other things, he was aware of the need for large numbers of observations. He also stressed the importance of taking into account the whole range of statistical variables, of making "proper allowance for all collateral circumstances that may have influenced the general results."[38]

Pliny Earle, who successively served Philadelphia and New York asylums, pointed out many ways for his fellow administrators to improve their statistics. He noted how difficult it was to generalize from American data about anything more complicated than the greater curability of females than males. And he went on to point out errors being made by his fellow superintendents in calculating such matters as the annual mortality in institutions and the "average" proportions of cures. He advised superintendents against attempting to draw numerical conclusions about the results of treatment until the institutions had existed for several years. However, Earle believed firmly in the value of comparative data, and his 1841 survey of asylums had used statistics from the reports of American institutions as well as those of the Europeans. Because these data varied widely, he hoped that American asylums would adopt "a common formula for the statistical part" of their reports. Their data would then be more useful in reaching conclusions about the nature, causes, duration, and curability of insanity.[39]

In Massachusetts, Luther V. Bell, of the McLean Asylum, in 1841 took a sharply different view of Earle's compilation of data from the American institutions. In fact, he thought that by and large the attempt at comparing the data of such institutions was useless, and that reports of them should be discontinued except for internal use. In the present state of knowledge, most reports were completely inaccurate, if only because there was no way to take the many variables into account. Bell criticized those who compiled such data for "attempting to give a numerical expression to incidents having no mathematical relations," and he argued that few of the really significant matters having to do with the causation, duration, or curability of insanity were "capable of being stated numerically."

> I presume that the ages of patients, their civil state as married, widowed or single, the color of the hair and eyes, the complexion and the like, might be recorded and conveyed with a tolerable degree of exactness, but beyond this any thing of the ordinary tabular statistics requires so many explanations and qualifications on account of their complexity, uncertainty and changeableness, as to deprive them entirely of the character to which they seem to aspire.
>
> When it is practicable by any acuteness of sagacity to seize upon, and by any precision of language to express in signs and classify in columns the degree and character of the affective sentiments, the moral qualities and the intellectual capacity of an individual, then we may look for the characteristics and results of insane hospitals in tables of scientific accuracy.[40]

Bell felt that even the meanings of the records in his own hospital were so uncertain that to publicize them was a disservice both to the community and to the medical profession.

Matters of opinion could usually be presented narratively with no risk, but when "published with a numerical aspect they may as *false* facts be of infinite injury." Bell endorsed Samuel Tuke's criticisms of statistics from English mental hospitals. Furthermore, he hoped that American physicians would organize to improve the use of statistics at home before the whole movement for care of the insane would be discredited by unfulfilled public expectations.

Isaac Ray, Superintendent of Providence's Butler Hospital, vigorously endorsed and furthered Bell's critique in 1849. Ray's appreciation of the necessity for precision had been honed earlier by his studies of the legal aspects of insanity. Now he found it incomprehensible that scientists would pass off their statistics of the curability of insanity as being scientific when only the most meager information about the effects of therapy was available. He discounted current attempts to discuss causes of the disease when there was a "total want of precision and uniformity in the use of language." He belittled the usefulness of classifications of types of insanity so long as "no two of them are alike." And he deplored the lack of any uniform standard for completeness and accuracy of asylum statistics. So far, Ray thought, American asylum statisticians fell ludicrously short of the rigorous methodology of Quetelet. To approach this, he said, they would have to learn that statistics really was not just a simple, if laborious, gathering and adding up of figures. On the contrary, it "implies something more than a process in arithmetic. It is a profound, philosophical analysis of materials carefully and copiously collected, and chosen with an enlightened confidence in their fitness for the purpose in question."[41]

The comments of Bell and Ray represented extreme expressions of a vague dissatisfaction that leading superintendents of asylums felt about their statistical reporting. These medical men were aware of the "incorrect data, defective registration, gross carelessness and other causes" which often made antebellum statistical tables and reports totally erroneous. Yet, few were inclined to abandon numerical methods in studying insanity. Even fewer were prepared to give up these means of influencing legislators or other benefactors. Most superintendents undoubtedly agreed with Amariah Brigham's reply to Ray that the current statistical efforts should be encouraged rather than disparaged: "Admitting then, their partial imperfection, yet we must also admit they have accomplished much good."[42]

Removing such imperfections by generally elevating the level of statistical methods used in preparing the asylums' reports seemed to be the proper answer. However confused or erroneous the terminologies of mental disease, and however absurd many of the tabulations of meaningless figures, asylum superintendents continued to consider statistics as the main hope of progress in their field. Still, few American psychiatrists of this period were in a position to contribute very much to the overall improvement of statistics that Brigham had in mind. A notable exception was Edward Jarvis.

The Statistical World of Edward Jarvis

Edward Jarvis, a Massachusetts native, was one of the most painstaking antebellum students of insanity, particularly of its social ramifications, in the United States. Never successful in becoming a superintendent of an asylum, he nevertheless had a broader view of mental illness than most superintendents. Never especially effective in general medical practice, he became a success at treating mental patients in his home, and he had time left over for his investigations. Sharing much the same set of moral and cultural preconceptions as Dorothea Dix, Samuel Woodward, Horace Mann, Pliny Earle, and other reformers, he also shared their enthusiasm for quantification. However, Jarvis went into the methodology and structure of reform much more deeply than most of his generation. As a result, he became one of the first acknowledged American authorities on the use of statistics for social and medical problem-solving.

As a Harvard undergraduate during the late 1820s, Jarvis was indoctrinated in the methodical outlook of Baconianism, Scottish common-sense philosophy, and scientific classification. Subsequently he became immersed also in political arithmetic, empirical social inquiry, and Parisian numerical medicine. Thus embracing and utilizing virtually all of the rational com-

ponents of current statistical thinking, he ultimately epitomized the statistical mind of antebellum America.

Jarvis made his earliest original contributions to statistics in
Louisville, Kentucky, where he lived during the late 1830s and
early 1840s. Upon his arrival in the Ohio Valley in 1837, Jarvis
found that there was as yet little medical interest in Louis's
numerical method.[43] However, medical conditions that needed
statistical analysis were as abundant as anywhere, and mental
disorders were prominent among them.

Jarvis quickly made himself knowledgeable about the extent
of insanity in the Ohio-Mississippi basin, especially in Kentucky. It soon became apparent to him that the West had
neither enough asylums nor adequate care in institutions that
already existed.[44] The Mississippi Valley in 1841 had a population of over five million, and, Jarvis found, a "lunatic population of more than four thousand," with only four scattered
asylums to take care of them. "These could not contain a tithe
of all, who might be subjected to their influence, and not a
fourth of these who could be benefited by them." It seemed
evident that additional public asylums for the poor were required in Indiana, Illinois, Missouri, Arkansas, and Mississippi.[45]

For Kentuckians, the most urgent matter was not the erection of a new asylum, though another could easily have been
filled, but improvement of the old asylum at Lexington. Built
in 1824 just before the optimistic concepts of treatment took
root in America, the Lexington institution had become an
anachronism, an affront to the age of moral therapy. As late
as 1841 Jarvis found that it had no resident physician and was
still run essentially as a repository for the chronically ill, with
little attention being given to their care or cure. His comparison of it with other, more up-to-date asylums was startling.
Despite its high fence, he found, the Lexington asylum had a
prodigious rate of escape or "elopement," 1 in 11 as opposed
to 1 in 166 at ten other American institutions. More seriously,
deaths in the institution averaged 26 percent of the total inmates, as opposed to proportions of from 4.6 percent to 14
percent at nine other American asylums. Finally, the rate of
cures was clearly the worst in the nation: less than 40 percent

of the recently afflicted and 10 percent of the old cases were being discharged as cured, in contrast to the 70–90 percent of new cases and the 20–40 percent of the old which the other asylums averaged.[46] Jarvis went on to argue that Kentucky was indeed "doing the least" of any state for the public care of the insane, and the statistical ammunition that supported this argument ultimately helped persuade the state legislature to do the work necessary to upgrade the asylum.

Just before Jarvis's move back to Massachusetts in 1842, the results of the first national enumeration of the insane were made public. Jarvis initially made a modest critique of these census data, merely pointing out the enumerators' failures to identify considerable numbers of the insane in many local communities. However, after examining the data more closely, he discovered that the errors were far more serious. Mistakes in tabulating the cases of insanity among Northern blacks had produced an enormous discrepancy in the apparent incidence of the disease from one region to another: the free states seemed to have a ratio of 1 case per 144.5 blacks, the slave states only 1 in 1,558.

Jarvis and other indignant Northerners quickly petitioned the House of Representatives to have the errors corrected. However, Secretary of State John C. Calhoun found the original census statistics far too useful to think of changing them. In fact, throughout the remaining antebellum years, he and other apologists for slavery frequently cited the reports as conclusive scientific proof of the beneficent effect of their "peculiar institution" upon the black.[47]

In Massachusetts Jarvis continued to be less interested in the clinical nature of the cases of insanity under his care than in the social ramifications of the disease. This interest resulted in his conducting several complex statistical studies during the 1850s. In one, he examined the accumulated annual reports of 250 hospitals for the insane in eight countries in an inconclusive effort to determine whether men or women were the more liable to become insane.[48]

Jarvis also attempted to answer the question of whether insanity was really on the rise, as most people thought it to be. As of 1850 no accurate answers to this could be obtained from official censuses, he noted, since no government in the world had yet conducted the successive reliable enumerations of its insane which would give the necessary comparative data.[49] Likewise, no adequate judgment could be formed from successive annual figures of asylum populations, because no asylum included all of the insane in a community. Even the enormous increase in demand during the past dozen years or so for admission to existing asylums was no indicator of an increasing incidence of mental disorders, though that demand was so great that the institutions had filled to overflowing and new ones had to be built in nearly every state. Jarvis thought this phenomenon was more accurately simply a reflection of the early successes of the asylums, the spread of public knowledge about the humane moral treatment of insanity, and the greatly increased public confidence in its curability. In consequence, instead of revealing statistical clues—"ascertained and enumerated facts"—about the trends of insanity, the rise in asylum populations was simply "an indication of the degree of civilization, and of general intelligence, especially in respect to mental disorder, of public generosity, and of popular interest, in behalf of the afflicted."[50]

These questions, together with such urgent concerns as the susceptibility of foreigners to insanity and its close relationship to poverty, were examined again at greater depth in Jarvis's 1855 study of insanity and idiocy in Massachusetts. There Jarvis had the opportunity not only to make an exhaustive survey of a limited population but to demonstrate the kind of statistical thoroughness which he had found lacking in so many other studies. He based his inquiry on the assumption that medical facts about insanity were best investigated by doctors rather than by census marshals. He did this by directing questionnaires primarily to the entire medical profession of the state, and he backed them up with inquiries to overseers of the poor, jail masters, hospital officers, and a few clergymen.

Out of the 1,319 medical practitioners of all varieties in the state who had been identified, 1,315 eventually sent replies to Jarvis, clearly a remarkable response to a questionnaire in any day or age. To some extent this impressive result may have reflected the spread of statistical interest among the Massachusetts medical profession, but it was made possible only by Jarvis's insistent cajolery and follow-ups. Once compiled, the report provided the Massachusetts legislature with the necessary justification for a new state hospital, and it subsequently served reformers of other states with ammunition for similar campaigns. Such reformers, together with his medical contemporaries, gave Jarvis the recognition he deserved for producing an important and influential sociomedical document. Most of all, however, they admired the technical expertise which made his report a model of statistical excellence.[51]

With other psychiatrists, Jarvis thought that public documents like his and the reports of the mental hospitals should be extensively publicized and made freely available. In 1857, on behalf of the Association of Medical Superintendents of American Institutions of the Insane, he compiled a list of 161 libraries in which the various hospitals were urged to deposit their reports regularly.[52] Even as he promoted the use of these institutions, however, Jarvis saw that the very demand for and existence of the asylums represented a colossal failure of society as a whole. Moral philosophy had been more or less successfully applied to the treatment of the insane in these institutions. However, the application of its principles to individuals outside the asylums had not really been systematically attempted, though logic suggested that such applications might well prevent a large proportion of potential patients from ever being afflicted with insanity, or, for that matter, with any other sickness.

During the 1840s Jarvis had become as dedicated as any phrenologist or health reformer to educating schoolchildren and the public in the basic laws of hygiene and physiology. With a few other regular physicians, he worked to turn the orthodox profession at least partly from its traditional preoccupation with curative medicine to the more humane and socially productive practice of preventive medicine.[53] Every individual, Jarvis explained, had a certain "capital of life"— an aggregate of physical and mental energies which produced what he termed "vital force." As with business capital, one's capital of health and life could diminish if certain basic rules were not followed, but it could be preserved and increased by faithful attention to food, dress, exercise, and other aspects of personal hygiene. "The laws of life are as fixed as the laws of matter," Jarvis went on. "The quantity of vital force which is generated, the amount of health, strength, and enjoyment, which is obtained, are in mathematically exact proportion to the fulfillment of the conditions of life."[54]

Jarvis came to think of sickness as one form of poverty. By careful attention to the laws of hygiene one could avoid this form of poverty, increase his vital force, and add months or years to his life. It was, of course, primarily the responsibility of each individual to follow the rules of basic hygiene. Still, because not everyone had the intelligence or self-discipline to do this, it also became a responsibility of the physician to keep track of the normal level or standard of vital capital possessed by each of his patients, "in order that we may compare the diseased state with it, and measure the extent of the error."[55]

Jarvis realized that society as a whole also had a responsibility in this matter. For many people, especially the poor and otherwise disadvantaged, the maintenance of vital force at a desirable level was impossible, because of conditions that were outside their capacity as individuals to modify. The state had accepted the mandate of the people both to count and to care for those whose vital force was diminished by a mental disorder. Jarvis and his generation of reformers now turned their attention to getting the state to do more about preventing illness in the community, whether mental or physical. Improvement of the state's methods for collecting vital data was the essential first step toward this goal.

8

Vital Statistics and Public Health

The numbers of America's insane were determined from enumerations or estimates of actual cases, but this was not true of most other diseases. In fact, the primary measure of a community's health at this time was its mortality, not its morbidity, and the extent of mortality was based predominantly on counts of burials rather than of certificates of death. Until the 1830s, most people's concern for public health was limited to the individual cities and towns in which they lived. By the 1840s, however, as part of a general expansion in the horizons of civic consciousness, many began to be concerned also with the health and mortality patterns of entire states. Since the bills of mortality of scattered towns obviously could not tell much about health and mortality in a state, the creation or renewal of systems for registering vital statistics became an increasingly urgent matter. The effort to establish such systems in the United States was both the first phase of an organized public health movement and a significant early stage in the professionalization of statisticians in this country.[1]

Several European countries had national registration systems by 1840, and physicians in a number of them—Louis Villermé and Alexandre Parent-Duchatelet in France, Johann Ludwig Caspar in Prussia, and others in Switzerland, Italy, and Scandinavia—were using the vital statistics to illuminate public health problems. Great Britain did not replace its ancient parish registers and bills of mortality as sources of vital data until 1836, when Benthamite poor-law reformers pushed through legislation for an effective nationwide registration system. Beginning in 1837, the annual reports issued by the

General Register Office under this system, and particularly the comprehensive statistical analyses by the Compiler of Abstracts, Dr. William Farr, provided an eye-opening picture of the excessively high death rate in a country swept up in the industrial revolution. Along with lurid details in a host of later reports on urban sanitation by Edwin Chadwick and other reformers, these reports did much to awaken Americans to the need for overhauling their own registries of vital statistics and measures for sanitation.

An Era of "Graveyard Statistics"

The usual sources of data for the bills of mortality in early nineteenth-century American cities were the sextons of the various cemeteries who reported, weekly as a rule, on the burials there.[2] Any consideration of urban health problems thus focused closely upon the work of the sexton; the numbers of burials became both medically and statistically important.

This necessary preoccupation with death accompanied its romanticization in other circles of that period in history. Poets like Bryant and Poe, along with painters like Washington Allston and Thomas Cole, emphasized the transitory nature of life and the gloomy finality of death. And the attraction of the phrenologists' old skulls was for many less a matter of science than a perennial fascination with the mysteries of death.

Romanticism and realism came together when city dwellers began trying to move cemeteries to the suburbs. As the cities filled up, the demand for land put pressure on churches to sell off their adjoining graveyards, which were rapidly running out of space for further burials. Moreover, because they were often poorly tended and unsightly, frequently gave off offensive odors, and allowed burial practices that were widely regarded as dangerously unsanitary, removing cemeteries from the crowded areas seemed increasingly necessary.

Observers pointed out that in older cities cemeteries constituted some of the most deadly sources of miasmas from decaying material, and hence were among the most concentrated foci of "febrile poison."[3] Some became concerned with such matters as the rate of decomposition of bodies and the maximum number of burials medically safe to permit in a given space and in different soils. Others were amused by the calculation that, if all of the dead had been evenly distributed around the world since the beginning of time, there would be one hundred bodies stacked up in every grave.[4]

Under various pressures, some churches did close their downtown graveyards and opened new ones outside the city. At the same time, beginning in the late twenties, the concept of the private "rural cemetery" took hold among the well-to-do. By midcentury dozens of such cemeteries with idyllic settings had sprung up in or near cities all over the country, about twenty around Philadelphia alone.[5]

Whatever the nature of their cemeteries, cities usually remained heavily dependent upon the burial information supplied by sextons until provisions for registering deaths could be adopted. However, other means sometimes had to be devised to obtain some idea of the extent and distribution of mortality. New Orleans offered an example.

When Benjamin Latrobe arrived in New Orleans shortly after the War of 1812 to supervise erection of a municipal waterworks, he found that no official bills of mortality were being issued in the city. He attempted to gather figures on burials on his own but quickly found himself frustrated, chiefly by the poor records of the Protestant churches.[6] However, within a quarter of a century, certain changes in burial practices in the city made a special kind of private mortality enumeration possible.

As Latrobe noticed, the poor of New Orleans were usually buried in the great Bayou cemetery of that city in graves that were quickly water filled and crawfish infested. Interment of the well-to-do, on the other hand, was increasingly in cemeteries with large and often imposing elevated tombs and vaults. The New Orleans physician Bennet Dowler in 1840 took advantage of the rather full inscriptions on these tombs to obtain some of the kinds of vital information that bills of mortality should normally have provided. Specifically, he obtained the mean ages from a sample of several hundred inscriptions in various cemeteries. Comparing them with similar data from Charity Hospital, Dowler concluded that New Orleans was an entirely salubrious community for long-time and well-to-do residents, but he conceded that it was highly unhealthy for newcomers, notably poor immigrants arriving during the hot season. Dowler, who had few peers among those he called "grave-yard arithmeticians," eventually applied his mortuary findings to the vituperative debates of the 1850s over New Orleans public health measures. Denying that the city really deserved its reputation as the "Necropolis of the South," he forcefully argued against the sanitary measures which had been proposed by the board of health largely on the basis of statistics of the poor from the Charity Hospital. Consideration of data from all classes was essential, he insisted, to preserve the good name of the city.[7]

Tombstone inscriptions were not satisfactory substitutes for official bills of mortality, and recognizing this, physicians of cities without regular bills pressed their medical societies to obtain the necessary local provision for them. Where bills were already issued, sporadic efforts were made to eliminate their imperfections. As early as 1826, Walter Channing of Boston outlined four basic specifications for these documents: (1) causes of death should be certified by physicians before any burial; (2) physicians should be honest and accurate in assigning causes; (3) a uniform terminology should be used; and (4) preparations of bills for publication should be done by a physician, perhaps the port medical officer. Other individuals pointed out the desirability of obtaining birth statistics along with those of burials, while some began to argue for statewide registers.

Given the squabbling between medical factions, most of these goals proved unrealistic during the 1820s and 1830s. In Boston, the local medical society was frustrated for a decade in its effort just to obtain a standard nomenclature for the bills. Though the city government accepted the society's nomenclature in 1837, individual physicians continued to submit masses of unusable and "chaotic materials" and paid no attention to the official list of diseases. Reports thus remained thoroughly jumbled in their terminology and limited in their usefulness.[8]

Despite their manifest shortcomings, there was recognition that the bills had at least some scientific value. Particularly in cities where bills were issued regularly, the accumulating data permitted increasingly detailed studies of urban sanitary conditions. By the 1820s and 1830s, a few individual physicians, characteristically those with experience working in dispensaries or on boards of health in the larger Eastern cities, were taking advantage of these accumulations of data on interment. Using such figures and other information, they undertook studies of disease and mortality patterns over varying periods of years. Their compilations were published in medical journals as "medical statistics," a term already used in European periodicals. In some cases, the analyses became rather comprehensive.

One of the earliest analyses sketched out ideas for municipal health administration and statistics that went beyond Channing's suggestions. In 1827 two young physicians, Nathaniel Niles and John D. Russ, published a short comparative study of New York, Philadelphia, Baltimore, and Boston mortality data for periods of from seven to eleven years. A few concise tables organized the data by disease and race and established the proportions of deaths to the population. The authors found the New York data superior to the others because they had been prepared "under the official superintendance of a medical gentleman," an arrangement which they recommended for other places: "Would not the ends of justice, as well as the cause of science, be promoted by the appointment of a medical officer in every city, whose duty it should be to examine the circumstances attending the death of every individual, and to make a return of the age, sex, profession, disease and its duration, as well as the time and place of death; regard being always paid to the circumstances of colour, and of freedom or slavery."[9] Niles and Russ thought that such officials should also devote attention to the effects of such factors as climate, topography, occupation, and economic position on disease. Standard forms for reporting data should be adopted to make comparisons between cities easier and more meaningful. Birth records were essential, and comprehensive social statistics were desirable if the overall health and degree of perfection of society were to be known.[10]

The Niles and Russ report was something less than a definitive blueprint for public health statistics of American cities, and it was also premature. Cities did not rush to engage medical officers and give them health-related statistical duties. Even New York did not consistently keep physicians in its office of City Inspector. But other individual physicians did undertake to expand on the types of public health data that this report had presented. In 1832, for example, D. Humphreys Storer prepared a somewhat comparable but longer analysis of the Boston bills of mortality for nineteen years. And in 1836, Charles A. Lee did a similar study of the New York bills for sixteen years. By that time, in fact, many physicians of the larger Eastern cities were expecting the bills, whatever faults they might have, to be published regularly and to conform at least to the norms of good visual presentation of data. A Boston editor was thus gratified to find in one of the New York bills, "a voluminous and melancholy document, extremely well executed in its tabular details." And a Philadelphia bill of this period, too, whatever its content, seemed to be "published in a very neat tabular form."[11]

Thanks to the meticulous Quaker physician Gouverneur Emerson, the Philadelphia bills were more systematically utilized and analyzed during the decades before 1850 than those of any other American city. From his work, first with the city's dispensary and then as member and Secretary of the Board of Health, Emerson acquired an intimate knowledge of the urban conditions which the bills reflected. In the latter post, he also became more familiar than most with the mechanics of compiling the bills. Beginning in 1824 he published a number of annual commentaries on the bills. Even more important, over the next quarter-century he prepared a series of detailed analyses of the bills for periods of ten or more years. These reports, more than any others, became the models for the quantitative health reports that subsequent generations of American sanitary reformers found so important.

For Emerson, statistics properly used provided "distinctness to views" and "accuracy to conclusions" in social, political, or scientific investigations. Well before the British health reforms got under way, he saw that, as used in public health, "by demonstrating the existence of evils they may lead to a removal of their causes, and serve as a test by which to determine the success or inefficacy of the measures resorted to for that purpose." He thought that the Philadelphia burial regulations ensured excellent data for the bills of that city. He also felt that the listing of causes of death on the bills was reasonably accurate; many of the Philadelphia physicians who certified these causes were accustomed by 1827 to checking their pathological opinions by making autopsies. All in all, the Philadelphia mortality data were as well recorded as those of most of the large European cities, it seemed to him.[12]

Wherever bills were issued regularly over numbers of years, the various medical analysts were able to point out patterns of endemic disease and mortality, but of course the patterns were broken from time to time by epidemics. Smallpox, for instance, began to flare up in some cities in the decades following the original enthusiasm for vaccination, because they had large new populations lacking immunity to that disease. During such outbreaks, interest in the bills of mortality invariably increased, and one or more physicians could usually be counted upon to gather supplementary data. These outbreaks were usually local and relatively small in impact, but that was not the case with a new and lethal disease that invaded the United States in 1832.

Cholera Record-Keeping, 1832–1854

The cholera, which struck after a relentless march across central Asia and Europe and down through Canada, arrived after the most advance warning any epidemic had ever had in this country. It stirred up greater fear than any pestilence since turn-of-the-century waves of yellow fever, and the feeling of powerlessness which rose out of the mounting death rates from the disease helped fill the churches on public fast days as they had been in the years of the yellow-fever epidemics. The regular medical profession, in its state of chronic uncertainty, could agree only that early treatment was essential.[13] Physicians tried the old therapies of bleeding and purging and almost any other measure that came to mind, including applications of stimulants, opium, calomel and other drugs, and saline solutions. None appeared to help much. In fact, hospital experience with the different therapies in this country and in Europe subsequently demonstrated "how little reliance the

profession can place upon any empirical or exclusive practice."
With patient mortality ranging from 47 to 90 percent for the
various regular therapies, it was no wonder that individuals
increasingly turned to alternative therapeutic systems.[14]

The cholera called forth an immense amount of discussion,
both lay and medical. A large proportion of this was carried
on in quantitative terms or supported by statistical data that
went far beyond the information available in bills of mortality.
In the medical press, the deluge of writing on the subject
started late in 1831, with historical accounts of the disease's
origins in India, reports of its progress in Europe, summaries
of pathological findings, and reviews of different modes of
treatment. As the epidemic reached France and England in
early 1832, and the mortality figures became more ominous,
American reporters began to vie with each other to issue the
most comprehensive and authoritative accounts.[15] When the
cholera finally reached the American continent, the rate of
publication accelerated tremendously. Reviewers who tried to
keep track of the outpouring of writings on the subject between 1831 and 1833 concluded that, medical journalism had
come of age even if therapeutics had not. The editor of the
Boston Medical and Surgical Journal maintained that his weekly
was the only one to read during such a crisis because of the
importance of getting "the earliest possible information respecting the march and mode of arresting that appalling malady." But monthlies, quarterlies, and books sold equally well
during the cholera outbreak. It quickly became apparent to
editors that "it will not be the fault of the present race of
physicians if posterity should obtain an inadequate idea of the
history of the present epidemic."[16]

In specific communities the official ad hoc responses to cholera included ineffectual quarantines, the creation and outfit-
ting of special hospitals, the cleansing of streets and houses,
and the gathering and dissemination of statistical information.
Since the official response to the epidemic was invariably inadequate, many private organizations added their energies to
the public effort. Together public and private groups produced a mass of data that were not always systematic but which
left clear pictures of the distribution and effect of the disease,

especially in large cities. Contributing to that picture in Philadelphia were daily canvasses of hospital, prison, and almshouse records, special daily reports from private practitioners on cases and deaths, and equally frequent tallies of interments.[17] In New York, at the height of the epidemic, "an association of physicians" published a *Cholera Bulletin* several times weekly, while Philadelphians issued the *Cholera Gazette*. Such publications suggested new therapies, passed on information from or criticisms of the Board of Health, and printed a variety of tables showing mortality in other cities and countries, numbers of hospital cases, geographic distribution of local cases, and deaths by street—and sometimes spot maps supplemented the tables.

Smaller communities were as badly hit as the large, but they usually did not have the same access to information about what was happening elsewhere. A Vermont physician hoped that the *Boston Medical and Surgical Journal* would provide the medium for pulling together a comprehensive "statistical account of the progress of cholera in America," to be compiled from the letters of readers across the nation. The editors were willing, but most readers were much too busy during the course of the epidemic to send in information. Much of the current data published in this journal during the epidemic, therefore, was culled from newspapers rather than from hospitals or other medical sources, and physicians considered the data "extremely vague and unsatisfactory."

It was only after the epidemic subsided that a numerically minded physician like Samuel Jackson could prepare substantial analyses of his clinical findings about the pathology of cholera.[18] Moreover, even in retrospect the records of many communities were inadequate to furnish accurate epidemiological information. In Cincinnati, Daniel Drake estimated that some 3,500 whites had fled the city, that there were some 500 deaths among the 25,000 remaining whites, and that the blacks suffered disproportionately, with some 45 deaths out of a population of 1,500. But he had to confess that those were only guesses. "The total number of deaths from the Epidemic, will never, perhaps, be accurately known. Mr. Morrison, the clerk of the township, after availing himself of all official returns,

and other practicable sources of information," could only produce a rough estimate.[19]

When cholera returned between 1849 and 1854, there was
much the same terror among the people and just as much
scrambling to establish ad hoc health committees and facilities.
Uncertainty concerning treatment was still widespread. There
was an even more massive outpouring of cholera literature
and cholera statistics, both public and private, than in 1832,
and the statistics had many of the same shortcomings.

The fact was, as the AMA's Committee on Practical Medicine and Epidemics complained in 1850, that nothing important could be gleaned from most clinical reports on cholera
because there were so few quantitative data in them. There
were "often no statistics of numbers, ages, sexes, colours, no
numerical ratio of attacks to recoveries, no proofs of the value
of remedies, beyond assertions and boastings." The Committee also found that where statistics *were* supplied, there was "a
want of uniformity in the mode of making reports, which
obscures, or even renders inaccessible the truth."[20] As a result,
one physician could report that with his therapy 1 out of 1.7
patients was cured, while another, using "the same remedies,
in the same dose, in the same town, and at the same season,"
cured only 1 out of 8. Similarly, in a review of seven American
works on cholera published in 1849, D. F. Condie concluded
that "no statistics are given sufficient to enable us to judge
of the several plans recommended, nor any statement of the
percentage of deaths." For that matter, the only substantial
statistical comparisons of cholera remedies that were available to American physicians up to 1860 came from European
sources.[21]

Using various combinations of the standard therapeutic
agents, the average regular physician considered himself for-
tunate if he got through a season of cholera with a death rate
among his patients of only fifty percent. As in 1832, the regulars' lack of success in coping with the disease was not lost
upon unorthodox practitioners who used the frightful mortality figures to promote their own systems among the public.
Leaders of the sects were confident that if all of their practitioners would keep and publish proper case records, the re-

sulting statistics would completely discredit allopathic medicine and lead to its rapid downfall.[22]

Much more informative than the regulars' reports on therapies were the running statistical accounts of the progress of the epidemic in the various cities, abstracts of which appeared in the lay and medical press alike. Thanks to the recent advent of the telegraph, few if any communities now were left in the dark about the extent of the mortality, near and far, and about its imminent approach. Thanks to the recent agitation of medical societies for better bills of mortality, somewhat more of the cholera data were drawn from the established reporting mechanisms of cities. But supplemental measures again had to be adopted, particularly to obtain figures on numbers of cases treated by individual physicians and on those in hospitals or other institutions.

Physicians in several places drew up statistical comparisons of the 1832 epidemic and the later epidemics, but few yielded significant general information about them. One exception was in New York, where William P. Buel concluded that the epidemic of 1849 was less virulent then the earlier one, thanks in part to changes in population distribution, but even more to the availability of abundant public water supplies following completion of the Croton reservoir and aqueduct.[23]

As in 1832, some thought they saw the hand of God in the epidemic because of its "mysterious and capricious character." But those who looked more closely at the statistics found better explanations for it. At least a few cholera cases and deaths could be found in almost every part of any given town, even in the wealthiest neighborhoods. But increasingly precise data, and sometimes maps, made it abundantly clear that most of the suffering occurred in "certain localities and particular classes."[24] For instance, it was no surprise to anyone who had read Lemuel Shattuck's census report of 1845, that the Broad Street section of Boston, packed exclusively with poor foreigners to a density of 626,000 per square mile, was precisely the spot that subsequently had the greatest amount of cholera in that city.[25]

At midcentury just as in 1832, of course, there were substantial gaps, statistical errors, and related problems in the records of the epidemic. One of the problems in determining the extent of the later epidemic was the large number of individuals who fled the cities with their sick or who buried their dead secretly without reporting them to the officials. Again, in many cities the reporting of cholera cases by members of the medical profession fell off drastically in the course of the epidemic. In places where physicians failed even to provide certificates of death, sextons had to report large numbers of cholera victims as having died from "unknown" causes. In Cincinnati, regular physicians refused to report any cases at all after an eclectic practitioner was appointed to the staff of the city's cholera hospital. The irregulars, moreover, were accused of reporting "everything as cholera, in hopes of extending their own reputation by spreading their marvelous histories of cures."[26]

A final difficulty was that reports on cholera from boards of health, like the bills of mortality, were often prepared by unqualified personnel. Doctors frequently found that such documents included a jumble of undigested data and generally were "very defective." It was not that the boards lacked large enough staffs during the epidemic, for the Philadelphia Board of Health was said to have enough employees in 1849 "to conduct the affairs of the South American Republics." The staffs were just not yet competent enough in Baconian fact-finding to make the study of cholera truly professional.[27]

Epidemics have often been credited with providing the spark leading to improved public health in nineteenth-century America, but in fact, their effects have been mixed. The long-range campaign for improved public water supplies was undoubtedly advanced by the fear of cholera, but the special provisions for cleaning the cities, the temporary hospitals, and the auxiliary data-gathering mechanisms only rarely became more than ad hoc arrangements, and most were dissolved with the end of the epidemics. The movement for better bills of mortality achieved permanent gains in some communities but elsewhere showed few immediate results following the cholera epidemics and had to be launched all over in the medical societies. The ideal of paid public medical officers to prepare the bills and look into matters bearing on community health on a continuing basis regardless of epidemics remained as far from reality as ever. As a result, the analysis of health conditions continued to be a sporadic activity of societies or, more often, of individual physicians working from inadequate graveyard statistics, while the direction of any public sanitation measures remained the responsibility of overworked aldermen or politicized physicians on part-time boards of health.

The chief impetus for constructive changes in the bills of mortality, as well as the main scientific arguments for other public health measures, unquestionably came from the regular medical community. This was true despite the increased involvement in matters of health by temperance lecturers, home missionaries, and other lay reformers. Divided as it was, however, by therapeutic differences, sectarian antagonisms, and competitive squabbles between medical schools, the medical community lacked the authoritative voice needed in political circles to ensure the changes, and its members generally lacked the statistical expertise necessary to shape them. No significant improvement in vital statistics could be reasonably anticipated unless some other organized body or group also interested itself in the problem. In the late thirties and early forties such support materialized as statistics emerged as a self-conscious profession forming its own institutions and working to improve American methods of applying data.

The Professionalization of Statistics

Statistics first appeared as an identifiable specialty in the Northeast, but some indications of specialism were soon evident in other sections of the country. During this early stage, individuals who had to deal with various kinds of data identified vital statistics as one of the crucial elements of political arithmetic needing attention in nineteenth-century industrial communities. Accordingly, almost from the first, the statistical movement involved itself in public health through its substantial scientific effort to improve or replace graveyard statistics by something better. Its campaign to obtain systematic statewide registration of such data thus directly paralleled and supported the efforts of the medical profession.

In the midthirties, intellectuals who thought about the subject at all generally agreed that if statistics were used, they should be used with care. Yale's Benjamin Silliman observed that the United States needed accurate, carefully analyzed data more urgently than any other country, "because we are still in the forming stage of society, because our interests are immensely diversified, and because in this republic . . . man, in high intelligence, is in a state of the greatest activity, with the most numerous and powerful excitements and with the feeblest restraints." Silliman thought that the federal government was the logical agent for obtaining these data: "It is certain that nothing but legislative enactment can ever give us a full view of the varied statistics of so vast a country as this.[28]

James Fenimore Cooper thought that, by and large, Americans were inherently "too wary and too ingenious" to fall into serious errors in their statistical calculations.[29] Others, though, were not at all sure of native ingenuity. On the contrary, the uneven quality of early nineteenth-century American statistics clearly indicated that there was much room for improvement. Moreover, because the aspects of national life undergoing statistical analysis—commerce, agriculture, occupations, manufacturing, crime, education, and many others—had proliferated, it became increasingly necessary, if data were to reflect the state of the nation accurately, for the compilers to understand and employ correct statistical procedures. As more individuals devoted larger amounts of their time to these statistics, the need for some form of technical guidance and professional exchange became more and more imperative.

The New York analyst Archibald Russell in 1839 noted four conditions that favored the cultivation of statistics in the United States: the broadly educated populace, the numerous local governments, the democratic insistence on full publicity for the operations of government, and the elaborate network of roads (over 200,000 miles) and post offices (some 12,000). In his volume on *Principles of Statistical Inquiry,* he did not, however, set out any professional guidelines for those engaged in such inquiries; instead, he confined himself essentially to arguing for the further broadening of the federal census.[30]

American texts designed for students of statistical science were actually several decades in the future, but professionalism was manifesting itself in a number of other ways: in the improvement of state and local statistical compendia or almanacs, in the increased use of ad hoc consultants to prepare census schedules, and ultimately, in the establishment of state bureaus of statistics, beginning with Louisiana's in 1849. Writers such as Gouverneur Emerson and J. D. B. DeBow began to identify themselves with or describe the "science of statistics," while editors in many fields increasingly filled their journals with articles heavy with statistical information.[31]

One of the principal manifestations of professionalism lay in the organization of statistical societies. Napoleon's abolition of statistical societies early in the century deterred European development of such bodies, but by the twenties, they were beginning to reappear in France, Germany, Mexico, and other countries. During the thirties, statistical societies sprang up in the principal British centers of the industrial revolution—Manchester, London, Glasgow, Liverpool, Bristol, and Birmingham—though some of these were short-lived.

By the thirties, Americans too were ready for statistical societies, and they were not far behind the Europeans in trying to organize them. But initial enthusiasms sometimes surpassed capacities. Charles Sanderson, a New Yorker, reported in 1835 that the European societies had already proved themselves and that similar bodies would be of immense value in guiding the growth of the United States: "the embryo gigantic powers of this Republic are now beginning to develop themselves, and it is of primary importance that the grand stream of prosperity be directed into that course which will not only secure the present prosperity, but also the future greatness of the United States." To provide expert counsel toward this end, Sanderson urged the creation in principal American cities of groups that would be modeled after and would cooperate closely with the Universal Statistical Society of France. However, when the New York Statistical Society was incorporated in 1836, it used the London Statistical Society as its model, with divisions devoted to economic, political, medical, and moral and intellectual, statistics. Since the necessary number of paid memberships apparently did not materialize, this society never actively met to conduct business. Accordingly, New York did not have any statistical organizations until the fifties, when the American Geographical and Statistical Society was formed in 1851, and the Society for Statistical Medicine appeared soon afterward. Meanwhile, the Statistical Society of Pennsylvania was organized in 1845, but it, too, became inactive within two years because of lack of participation.[32]

The first American society in the field that had staying power, and certainly the most influence socially and professionally, was the American Statistical Association. Founded in Boston in 1839, and again modeled largely on the London society, this association aimed at "collecting, preserving, and diffusing statistical information" in virtually every area of knowledge. It also proposed "to promote the science of Statistics, to suggest and prepare the best forms for keeping records, proposing questions and making investigations."[33] Despite the national scope implied by its name, as well as by its lists of honorary, corresponding, and foreign members, the association was essentially a local society and remained so until long after the Civil War. But this concentration on largely local affairs turned out to be extremely fruitful in the long run, both for statistical professionalism generally and for the advance of vital statistics and public health in particular.

The association's influence in Massachusetts was due in part to the variety and prestige of its membership. Though it included prominent businessmen, it also included scientists like Benjamin Pierce, George P. Marsh, and Joseph Worcester. Along with prominent clergymen and lawyers, there were also educators, members of the legislature, and state officials. The ranks of social reformers were represented by Samuel Gridley Howe, Horace Mann, and others.[34]

Medicine too was well represented. Among the earliest members were such Paris-trained clinicians as Elisha Bartlett and George Shattuck; physicians like Luther Bell and Edward Jarvis, who were involved in the care of the insane; and general practitioners such as John D. Fisher, David H. Storer, and Augustus A. Gould. All in all, fourteen of the fifty-four original local members of the association were physicians.[35] This proportion of doctor-members helped ensure the body's concern for health-related statistics. It also facilitated cooperation between the association and local medical institutions, both in gathering such statistics and in furthering common legislative ends.

As so often happens, such activities depended upon a small number of particularly motivated members. In fact, during its first decade a nonphysician, Lemuel Shattuck, stood out in the formulation and pursuit of many of the association's health-related objectives. At the same time, Shattuck was carrying out his own statistical work so competently that it became a professional standard for the rest of the association and for the public at large.

Lemuel Shattuck—"Statist"

Shattuck's early career was spent chiefly as a storekeeper, bookseller, and publisher in Concord, Cambridge, and Boston. Active in Massachusetts local history and genealogical circles, he published in 1835 his first large work, *A History of the Town of Concord*.[36] In his pursuit of authentic materials for the book, Shattuck became aware of serious gaps in the historical record of Concord, particularly the details of births, marriages, and deaths. Some bills of mortality could be found in the town clerk's office, and the venerable Ezra Ripley, pastor of the First Church, possessed a volume of early church records. However, these were far from enough. In fact, the shortcomings led Shattuck to realize that obtaining a workable state system of compulsory vital statistics registration would be the only way to ensure having such data for future histories.

After Shattuck moved to Boston in the midthirties, it became increasingly apparent to him such registers could be useful far beyond their service to history, genealogy, or to the legal claims of individuals. In his varied service to the city and state governments, he came to see this larger role for vital statistics as a part of political arithmetic, and notably as a key to improvements in public health. Shattuck's efforts to expand this role epitomized those of the professional "statist"—the statistically oriented analyst emerging in the forties and fifties to seek the enlargement of governmental bodies and mechanisms to obtain greater amounts of data to apply to social, economic, and other problems.

Shattuck's Boston career as a statist was launched from his bookstore. Like the better-known bookstore of his contemporary, Elizabeth Peabody, Shattuck's establishment, while it lasted, was one of the focal points of creative change in the Hub. In her store, Miss Peabody, in addition to distributing tracts for women, provided a focus for Romanticism and an active center for the spread of New England Transcendentalism. Shattuck, by contrast, ran a shop which, at least in retrospect, stood as a symbol of emerging statistical professionalism. There, at his desk, he drew up the original protocols of the American Statistical Association. He launched a wide correspondence, domestic and foreign, to solicit statistical reports, examples of registration legislation, bills of mortality, and related documents. And, using the bookstore as a base, he put his talents for ordering, numbering, and statistical analysis to work in a variety of ways in the interest of good local government.

Membership on the Boston City Council between 1837 and 1841 provided Shattuck with a far broader outlet for his energies than he had found in Concord. A fiscal conservative, he voted consistently for reduction of the public debt, and he resisted proposals for expensive civic undertakings, such as a public water supply. At the same time, he was ready enough to approve moderate expenditures if they promised to improve the existing mechanisms of government. Accordingly, he took on himself such innovations as devising a plan to collect and publish city documents annually, organizing a city reference library, and preparing an annual register of municipal officials and ordinances. However, all of these were subordinate to Shattuck's efforts to make the official vital statistics for Boston and Massachusetts more than the perfunctory records they had been.

Shattuck's position on the city council gave him leverage in asking foreign officials to donate statistical material for his studies. When he asked Quetelet, late in 1839, for Belgian statistical forms, reports, and a copy of *Sur l'homme,* he wrote of his aim to introduce an improved form of registration system into the United States and generally to obtain comprehensive statistical data: "I have the first subject now under

consideration in a committee of the City Council of this city."
In exchange for Quetelet's materials, Shattuck sent copies of
seven works which he presumably regarded as representative
of the best American statistical efforts of the 1830s. These
included his own *History of Concord,* abstracts of the early Massachusetts public school reports, a report of the Prison Discipline Society, the *Boston Almanac,* data on Massachusetts banks,
J. V. C. Smith's *American Medical Almanac for 1839,* and the
New York City bill of mortality for 1838.[37]

A search of the existing records of vital statistics in Boston
confirmed Shattuck's Concord findings: the past records were
grossly deficient and the official provisions for obtaining current data were ineffectual. He nonetheless painstakingly studied what scattered data had managed to survive and published
his findings in the *American Journal of Medical Sciences* in 1841.
Since the records before 1800 were quickly summarized, most
of Shattuck's article dealt with bills of mortality since 1811,
which made it of considerable contemporary medical interest.
However, its most important aspect was not its scientific content but its constructive suggestions for obtaining better data
in the future through state legislative action.[38]

By all accounts, the passage of the Massachusetts Registration Act of 1842 owed a good deal to this article, as well as to
Shattuck's other initiatives. Shattuck helped prepare drafts of
the proposed legislation, obtained sponsors in the legislature,
and rounded up endorsements from the medical and scientific
communities, including the Massachusetts Medical Society, the
American Statistical Association, and the American Academy
of Arts and Sciences. Once the bill was enacted, he furnished
the secretary of state with background documents and other
material to fill out the first registration reports. He also suggested much of the specific format of the reports, including
a modified form of Farr's classification and nomenclature as
a framework for the analysis of causes of death. Finally, his
written critiques, to the secretary of state, of the early operation of the registration system ultimately became the basis
for several amendments to the 1842 Act. For Massachusetts,
Shattuck fashioned a registration system that would be practicable and enforceable in local communities. He also made

certain that data brought together by the system would be properly analyzed, published, and made available for social and scientific purposes. The Massachusetts system not only survived its formative years but, by steadily improving, set an increasingly high standard for registration systems soon established in other states.[39]

Shattuck's campaign for registration in Massachusetts was systematic and efficient, as befitted an orderly man of commerce. And he hastened to assure the Massachusetts public that, unlike some other advocates of causes, he was far from being a visionary: "We are not a theorist . . . We are a statist—a dealer in facts."

At the same time, the immense new power of such facts was enough to turn a person into an enthusiast, if not actually a visionary. Shattuck realized that, through facts, people now had the power to increase the potential of their physical existences, much as Emerson argued that through Transcendentalism one could expand the horizons of the soul. Collections of vital data provided a means for accurately projecting the average duration of life and for constructing actuarial tables. They could furnish philosophers and political economists with information about the patterns of human life. They were useful to jurists, geographers, historians, statesmen, and reformers. And they could illuminate, for physicians and statists, the effects of external influences on sickness. By helping to lead to the reduction of sickness, they could presumably promote morality, wealth, and happiness generally. Shattuck had no doubt that registration data could answer some of the crucial questions of the times: Were the cities really less healthy places than rural areas? What were the long-term effects of typical American habits—the mobility, thirst for wealth, love of luxury? Did such habits "tend to check the progress of population, increase disease, and weaken the race?" What were the costs to human health and life of becoming a manufacturing nation?[40]

As one of the nation's foremost statists, Shattuck exerted considerable influence during the 1840s on the movement to make governments more responsive to human needs and purposes. With some of his contemporaries, he thought that, to date, governments in the United States had been excessively shaped by the demands of commerce. They "have legislated for property, but not for life. They have cared for the lands, the cattle, the money of their constituents, but not for their health and longevity.[41] Indeed, the large governmental expenditures for transportation routes, exploration, and geological and zoological surveys, were primarily motivated by business considerations.

Government action more equitably directed might not, of course, keep any given individual from disease or death. What it could do, at least, Shattuck pointed out, was to gather information that could be used for the benefit of all: "It is as certain that human life may be prolonged by knowledge and care, as it is that an ox will fatten, silkworm spin its thread, or a plant thrive, better, where knowledge and care are bestowed, than where they are not. Let the facts which the Registry System proposes to collect concerning Births, Deaths and Marriages, and the circumstances which attend them, be collected, digested, arranged, published and diffused annually, and their effects on the living energies of the people would be incalculable."[42]

The Physician as Statist

Shattuck thus had assumed the role of a philosopher of vital statistics, as well as its organizer, leading technician in this country, and salesman. But vital statistics was now predominantly a medical subject, one whose further development demanded a large and active role on the part of the physicians. During the next decade or two, the medical profession proved willing to become increasingly involved in this subject; in fact, the Baconian impulses of American physicians seemed to come together in an outpouring of statistical enthusiasm.

Some, like the New York physician Franklin Tuthill, were carried away by the vision of the vital statistician controlling the energies of the age and guiding the destiny of the republic. As navigator of the ship of society, the new statist had tedious labors to perform. But to Tuthill, this work, the labor of abstracting and analyzing masses of collected data, was the work of a scientist.

This quiet man will pour upon [his data] the strong acids, one after another, of his cogent reasoning, till all the irrelevant and impertinent are dissolved out, and the precious metal remains alone in its purity . . . From these leaden forms of statistics it is not the light rod of genius but the Lydian stone of patient study that receives the mark of truth. We may confidently expect them to reveal no less profitable secrets than those which we know are latent in our abundant meteorological tables, geological maps, and astronomical observations.[43]

While not all physicians became statists or had as grandiose a concept of vital statistics as Tuthill, a considerable number actively promoted their use among their professional colleagues. In Massachusetts, Edward Jarvis quickly became the state's most energetic and knowledgeable medical supporter of Shattuck's proposals for registration and public health legislation. Already familiar with English and Continental publications on these subjects, he wrote long critiques of the early registration reports of several of the American states. As longtime president of the American Statistical Association beginning in 1852, Jarvis became, after Shattuck's retirement from public life, the exemplar of the midcentury statist.[44]

Some of the more effective exponents of better medical and vital statistics were editors of medical journals, prominent among them, J. V. C. Smith of the *Boston Medical and Surgical Journal* and Bennet Dowler of the *New Orleans Medical and Surgical Journal.* However, none of the editors demonstrated greater intensity in their statistical enthusiasms than Samuel Forry.

Forry appeared, meteorlike, in American medicine about 1840, dazzled the profession for four years, and then died. An obscure young army surgeon, he originally achieved notice as the compiler of a massive statistical report on disease and mortality in the army covering the twenty years following 1819, when systematic records were first kept.[45] After resigning from the army in 1841 Forry began statistical studies of various aspects of health and disease in the general population. In 1843 he launched *The New York Journal of Medicine.* He filled that periodical's pages with every kind of article dealing with medical statistics, while through his book reviews he evaluated the statistical efforts of other physicians. He also began preparation of a book-length study of vital statistics, which he had virtually completed before his untimely death in 1844. For the next six years, several medical friends—chiefly Charles A. Lee and Austin Flint—tried to obtain a publisher for Forry's manuscript, but without success.[46]

Although the medical profession failed to find a way to get this theoretical work on vital statistics published, it made other contributions to the specialty, notably, to the spread and improvement of state registration. In Massachusetts the regular physicians set a pattern for the entire profession by not only supporting Shattuck's legislative proposals for a state registration system but by assuming a continuing responsibility for the operation and improvement of the system. The state medical society promoted registration through its committees, its policy statements, and its petitions, and a number of doctors took on professional duties for the state registry office. Most crucial were those physicians engaged by the Secretary of State as consultants to analyze the mortality returns and to arrange them according to the causes of death. Augustus A. Gould performed these labors for several years, followed by Josiah Curtis and Nathaniel B. Shurtleff. Like Shattuck, these early physician-advisers were impressed by the well-established registration systems of countries in western Europe, but they shaped most of the details of their new system, including the nosological arrangement, after the more recently established British plan. And, just as the statistical reports of the Registrar General of England had become medically important and useful because of the painstaking analyses of the physician and compiler of abstracts William Farr, so the Massachusetts statistical reports, under the guidance of other physicians, evolved into important medical documents.[47]

Like similar volumes, the Massachusetts registration reports were not the easiest kinds of reading. As Bennet Dowler observed: "Numerical Tables constitute a species of literature, the least attractive, yet discovered. Whoso wisheth to do scientific penance, let him read 100 pages of the [Massachusetts] figures." Nevertheless, Dowler and his fellow reviewers concluded their critiques with a "Well done, Massachusetts!" for that state's reports were indeed landmarks in the progress of American medical literature.[48]

Massachusetts provided a model not only in its legislation on vital statistics, its registration system, and its reports, but also in the close involvement of its physicians in making the system work well. In other parts of the country individuals soon set out to emulate the Massachusetts initiatives by getting their local medical societies to consider the question of state registration, and by petitioning their legislatures for action.

The creation, during the midforties, of a national medical organization was particularly timely for the spread of registration. The medical conventions of 1846 and 1847, and the American Medical Association which grew out of them, were concerned with such fundamental matters as licensure, ethics, the quality of medical education, and the problems posed by quackery and competing medical sects. Many of the participants were also interested in the advancement of medical knowledge. Since the gathering and use of statistics had become one of the significant, if not actually the principal, means of medical inquiry by midcentury, many delegates were interested in making vital statistics one of the new organization's official concerns.

At the first national medical convention, held in New York in 1846, the New York physician John H. Griscom led the move to make vital statistics a concern of the organization. Griscom, who had been exchanging views with Shattuck, had had experience with vital statistics during his brief tenure as City Inspector. Since 1843 he had been trying to obtain a registration act for New York State. Now he persuaded the convention to establish two committees.

The first, an ad hoc Committee on Registration, was composed of Griscom, Gouverneur Emerson, Alonzo Clark and James Stewart of New York City, and Charles A. Lee of Geneva, New York. Their report, adopted by the 1847 convention, provided for the convention to formally petition every state government to enact effective registration legislation and to request state and local medical societies to take the lead in lobbying for such laws. The convention then created a standing committee to follow through on the subject at the successive meetings of the AMA. Initially, this standing committee included Griscom, five other physicians—Emerson, Clark, Lee, Richard D. Arnold of Georgia, and John D. Russ of New York—and the layman Shattuck.[49] For the next fifteen years or more, these men and their successors kept the AMA actively engaged in working to obtain state registration systems that were more or less uniform, while other bodies, such as the American Association for the Advancement of Science and the National Quarantine and Sanitary Conventions, also joined in supporting that objective.

The second committee established at the 1846 convention resulted from a suggestion Shattuck made to Griscom for a group to prepare an authoritative nomenclature of diseases for use by the registrars. Use of the same terms was absolutely imperative if public health work was to be effective. Shattuck pointed out that these terms "are the measures and weights—the instruments by which the computations are to be made. Without such a uniform standard of comparison no just conclusions can be drawn. It would be equally proper to use Fahrenheit's thermometer in one place and Reaumur's in another, to estimate the comparative temperature of the atmosphere [as to continue to use different names for the same disease]."[50]

Griscom proposed that the committee on nomenclature include Emerson, Jarvis, Beck, and Lee, with Shattuck to be chairman and guiding spirit. However, the convention balked at having a nonphysician as chairman and named Griscom himself. Griscom "was compelled to accept this, in order to save the whole thing," but he prevailed on Shattuck to do the main part of the committee's work anyway. As a result, its report, as accepted by the 1847 convention, was very largely prepared by Shattuck. In it, the committee outlined means which, if followed, would ensure that the medical core of all of the anticipated data from state registration systems, the analyses of causes of death, would be prepared in a uniform and accurate manner throughout the country. To accomplish this, it proposed the adoption in the United States of a slightly modified version of the nomenclature and classification of disease used by Farr in the British registration reports. It also suggested a standard death certificate. Finally, in place of the astounding number and variety of names used on American bills of mortality up to that time—many of them vague, obsolete, or merely quaint—the committee urged that doctors work toward medical consistency by limiting themselves to a choice of 107 names for diseases.[51]

The convention's endorsement of this nomenclature and classification was a major contribution to the midnineteenth-century effort to attain greater precision in medicine. Getting individual members of the profession to conform to a standard terminology, however, was another problem. This became, in fact, a standing educational objective of medical societies at every level for the next seventy-five years, while the health officers and registrars, who hoped to extract meaning from the data of disease and death also helped in the effort. After the midforties, the number of these officials slowly grew as other states set out to duplicate the registration system of Massachusetts.

Antebellum Registration Difficulties

The AMA's recommendations on registration stirred up interest and action in many parts of the country. State and local medical societies began intensive efforts to persuade legislatures that registration was important. New York finally enacted a statewide registration law in 1847; New Jersey and Connecticut followed in 1848; Rhode Island and New Hampshire in 1849. During the fifties, Kentucky, Pennsylvania, South Carolina, Ohio, Vermont, and California also enacted such legislation. But these successes were deceptive. Even where legislation was adopted, it did not always work. In New Jersey and Rhode Island, nothing came of the early laws until amendments or new laws made them workable. In New Hampshire the law proved unworkable even when amended, while in Pennsylvania, after it proved impossible to gather data and issue reports, the registration act of 1852 was repealed (in 1855). In New York, returns were collected and two annual reports issued, but the registration system fell into disuse after 1849. Because of many operational difficulties, out of twelve state registration systems launched during the forties and fifties, hardly half were still functioning effectively in 1860.[52]

And not one, not even the Massachusetts system, was yet considered to have reached a level of excellence in which the data it collected were as complete or accurate as those collected under English or Continental systems.

Various technical factors contributed to the failures of registration systems: the attempt to get by with voluntary provisions; the designation of inappropriate local collectors or registrars (several states had disastrous early experiences in their use of school clerks); the failure to make sure the local registrars were paid adequately; and the absence of effective penalties for nonperformance of duties. But according to Jarvis there was an even more basic obstacle to the creation and continuance of the systems. In the mid-1850s, according to his view, the registration movement had moved too far ahead of its base of community support. It was a reform that had been perceived and shaped mainly by the physicians: "[But] these men are in the advance of the law-givers, and these are in advance of the people. In America, where all power primarily and ultimately rests with the people, it is useless for the legislature to adopt any law unless it is an exponent of the popular will already existing, or finds a response to it, and a support for it, in that popular feeling after it is adopted."[53]

While autocratic European governments could readily impose registration systems on its citizens, in the United States the electorate and its representatives had to be educated and convinced of the need before much progress could be made. Before the Civil War this often proved impossible. In California, the registration law of 1858 was jeopardized from the start by an "insurmountable prejudice against its provisions on the part of the people." In Georgia, Richard D. Arnold, who was president of the state medical society and a member of Griscom's AMA committee on registration, went to the state legislature in 1847 confident that he could persuade it to adopt a registration law; he was greatly chagrined when the legislators merely laughed at the idea. When another unsuccessful effort was made in that state in 1850, the opponents responded similarly, charging that the bill "was a trick of the Doctors." Even in Massachusetts, Shattuck and the early supporters of registration were so unsure of their popular and legislative support that they settled for admittedly inadequate procedures at the beginning rather than press for a better system which probably would have been rejected or later repealed.[54]

Progress toward registration in several states was halted by the outbreak of the Civil War. Several reasonably well-established systems, moreover, such as those of Kentucky and South Carolina, were forced to stop operating at that time. With the early momentum broken, the process of education had to begin virtually from scratch after the war. Consequently, it was many decades before adequate registration, even of mortality alone, could be realized in all of the states.[55]

There were other basic obstacles to the spread of registration. In not a few cases it was a simple matter of priorities. An AMA committee reported in 1856 that "the minds of the leading men in this country, as well as of the masses, are so much absorbed in commerce and the arts, agriculture, and politics, that it is almost an impossibility to turn them aside" to consider registration. Probably the most frequently expressed objection to registration was that it had no practical utility. A New York legislator derided that state's system as "a vain and useless incumbrance, an unprofitable tax upon our time and property." Others denounced it as a "costly luxury," and as an activity whose operation would introduce a repugnant degree of "espionage and inquisition" into a democratic society.

Such attitudes were clearly related to the popular distrust of elitist scientific and intellectual activities that James Jackson and Tocqueville had noticed. At the same time, there was a deep-seated dichotomy evident between rural and urban interests. Since rural legislators were not personally involved in the growing health problems of the cities, they felt no special urgency to consider the request of urban physicians for registration systems. Furthermore, registration was feasible only in states with relatively compact, concentrated populations. A Maryland commentator perceptively summarized the situation in 1850: "In a state possessing but one large city, and in which the country districts are, in general, far from being densely populated, many years will probably elapse before the State Legislature shall be ready to organize a general and complete system of registration."[56]

Meanwhile, those states that were successful in setting up registration systems properly emerged as early centers of statistical professionalism. Following the lead of Massachusetts, most states assigned the responsibility for gathering and publishing information under the new systems to the Secretary of State. In Kentucky, the Auditor of Accounts oversaw the system; in Connecticut, beginning in 1854, it was the State Librarian. In most cases, the responsible officials had no special training for these duties and inadequate time to perform them. Outsiders, usually physicians, were therefore engaged to collate the masses of data, analyze them, and prepare the statistical reports for publication. Sometimes, as in Massachusetts, Kentucky, and South Carolina, the labor was done by individual physicians working alone; in other states, such as Vermont and Rhode Island, the work was entrusted to committees of the state medical societies. Gaining expertise with several years of such work, several of these physicians—Robert Gibbes, Jr., William L. Sutton, Joseph Mauran, Thomas H. Webb, Charles L. Allen, and others—thus became members of what might be considered America's first identifiable generation of statists. And some put that expertise to work directly in the interest of the public health.

9

Sanitary Fact-Finding in the City

Antebellum statists soon became aware, from bills of mortality, vital statistics registers, and census reports, that a third of all deaths were preventable. Excessively high death rates were especially evident in urban areas for typhoid, dysentery, and various other so-called zymotic diseases; and the incidence of these diseases seemed to decline greatly when unsanitary conditions in a community were eliminated. The number of deaths from these preventable diseases at any given time came to be regarded as the sanitary index of the locality. John Griscom explained: "They are the hygienic barometer, whose figures on the scale denote the state of physical health, derived from the modes of life, the character of the dwellings, the condition of the streets, the attention given to the removal of filth, the extent and perfection of the sewerage, and the degree of intelligence and supervision of the health officers. The higher the degree, the more degraded are the people . . . judged by that standard."[1]

When the data were adequate, officials and physicians could determine from the sanitary index how a city's salubrity compared with that of other cities at home and abroad. They could also pinpoint the city's wards or precincts that had the highest death rates. Statistics provided the scientific foundation for community efforts to eliminate excesses of deaths over acceptable norms. They became the essential base of a slowly growing movement to systematically and regularly clean the cities in the interest of health. Statists rang the alarm for such reform, and they stood ready with the scientific expertise to direct it once it got started.

Sanitary Reform in England and America

Sanitary change in the antebellum United States was stimulated by many forces. However, as a self-conscious reform movement in the 1840s and 1850s it got its primary impetus from the dynamic, numerically based, and politically effective sanitary movement which had begun in Great Britain about a decade earlier. American enthusiasts took over much of the rhetoric of the British movement; for them as well as for the English sanitarians Edwin Chadwick, Southwood Smith, or William Guy, the sanitary question became "the great question of the day."[2] However, before 1860 there was little sanitary progress in the United States, and the ranks of the reformers were still small. Moreover, except during epidemics, reformers could not really convince most of their countrymen that there was any great urgency for sanitary action, as there was in Great Britain. To be sure, American cities were growing rapidly, and they had their growing pains, but most cities were smaller than the great British urban centers, and so far, all had experienced much less of the crushing impact that the industrial revolution was having on British cities: less of the crowding of populations near the factories; less of the intolerable strains on housing, transportation, water supplies, and sewage disposal systems; less of the dehumanizing concentrations of filth; less of the economic inequities and grinding poverty.

Nonetheless, the United States had its own urban sanitary problems, problems that multiplied alarmingly with the rise of native factories and technologies and with the crowding of poverty-stricken immigrants in slums. In response, Americans began early to steep themselves in the literature of the English (and to a lesser extent the Continental) sanitary movement. Particularly impressive were the reports of the comprehensive sanitary investigations of Southwood Smith, James Kay-Shuttleworth, Neil Arnott, Edwin Chadwick, and other zealots. Because of them, many Americans came to know the grisly details of conditions in London, Liverpool, or Birmingham, and what was being done about them. Moreover, through these British reports, vast collections of statistics which proved the value of sanitary works became available in the United

States to further the reformers' arguments before city councils or state legislatures for public health legislation.

Persuasive though they were, these remained British statistics pertaining to British sanitary efforts. A crucial reason for the lagging in the American public health reform efforts of the 1840s and 1850s was the failure of reformers here to organize adequate means for unearthing native data. American statists understood as well as anyone that to be useful as urban sanitary indexes their mortality data had to be so accurate that they could "be relied on as perfectly as figures in the adjustment of a business account."[3] Such precision could not be expected until American registration systems and public health surveys came up to the quality of their British counterparts.

Meanwhile, the statists admired and used the British information as best they could. They found great merit in the orderly, planned approach of the British reformers, even if it was predicated upon an unusual role for government. Lewis Rogers summarized the unique aspects of this methodology nicely:

> The sanitary movement of modern time, and of the present century, differs from that of antiquity in its more minute, thorough, and systematic character; it is founded upon a basis of statistical facts and philosophical principles; it recognizes and leans upon, as its main pillar of strength and support, the invariable and immutable relation subsisting between cause and effect; it works and projects its measures by rules and laws carefully deduced from a large mass of accurate and reliable observations.[4]

Like their British contemporaries, not a few of the early American sanitary enthusiasts thought of themselves as followers of Jeremy Bentham.[5] One aspect of utilitarianism that appealed to them as statists was the Benthamite passion for precision in government, and particularly in the orderly unfolding and operation of the general registration system. British sanitarians could use the Registrar General's early reports to great effect because they were systematic, complete, and regular, and because they included the remarkable analyses of William Farr. Americans found that the value of these documents increased "every successive year in a geometrical ratio." Americans thus quickly accepted Farr's formula of a death rate of 2 percent per year, or one death per fifty persons, as the acceptable rate for a normal community of that period. This yardstick allowed local antebellum reformers to measure the salubrity of their own communities and to set up specific goals for public health improvement.[6] In another report Farr confirmed as fact the belief that disease and death among city dwellers greatly exceeded that of country residents. Some Americans welcomed this confirmation of their ancient prejudices against cities; on the other hand, those who lived in cities and had to cope with urban problems found immense encouragement in Farr's conviction that the disease and death excess among them was preventable by sanitary measures, a concept which they saw translated in England during the 1840s by Edwin Chadwick into a dogmatic and effective formula for improving public hygiene.[7]

Another source of American interest in the Benthamite concern for the "greatest good of the greatest number" lay in the problem of poverty in British society. The poor-law bureaucracy under Chadwick provided England with an institutional base for coordinated national inquiries about poverty that were utterly out of the question in antebellum America, given this country's imperfect federalism. But the reports emanating from these inquiries provided Americans with endless instructive detail on the close interrelationships of poverty, filth, and excess mortality in modern industrial cities. Chadwick himself provided one of the principal documents with his massive 1842 *Report on the Sanitary Condition of the Labouring Population of Great Britain.*[8]

Charles Dickens, in the course of his American trip later that same year, urged every local official in the United States to study Chadwick's report.[9] Americans did read Chadwick with great interest. To be sure, some took umbrage at some of his data, particularly his figures suggesting that the conditions and duration of life in certain American towns were on a level with those of Liverpool, one of the least healthful of the English cities. But more saw in the report justification of their efforts to uplift society and health in this country and a model for their own sanitary investigations.[10]

To gain strength and stability, the American public health movement, like the British, required the interaction and participation of many people from diverse fields. Besides local officials and statisticians, these included social and medical reformers, physiologists and chemists, engineers, clergymen, educators, and temperance advocates. It especially included those who had anything to do with the urban poor—those who had experience with tract societies, almshouses, prisons, schools for the handicapped, asylums and dispensaries; those familiar with the education reports of Horace Mann, the hygiene tracts of William Alcott and John Bell, the data of the Prison Discipline Society, the bills of mortality of Philadelphia or New York.

For the middle and upper classes, urban sanitation became increasingly popular as they realized that it could greatly assist their overall efforts to elevate or cope with the poor. The movement for better sanitation began to gather force when reformers adopted it as the key to many of their other philanthropies. They noted its intimate relation to better public education, the spread of religion, the furtherance of temperance and morality. They realized that a clean physical environment was essential to the success of their campaign for individual hygiene. Statistical tables confirmed what everyone knew about the worst parts of any city, that those with the meanest dwellings, narrowest alleys, highest and most offensive piles of filth were those inhabited by the poor. Customarily these had been the last sections to get any sort of sanitary services, whether private or public. Now logic seemed to dictate a reshuffling of social priorities so that the undeserving as well as the respectable poor could get these services.

For some, the commitment to effective public sanitary measures came only after statistical demonstration of their need. John Bell, however, noted that, well before midcentury, statistics had already reduced to "poetic fiction" one of society's long-held delusions—that the poor neither needed nor wanted attention, that they were already happier, healthier, and longer-lived than the well-to-do. Villermé had amply proved the opposite for France. Physicians and statists had no doubt that American facts would show the same thing, and they proceeded to gather them. For such men, calculating the duration of life became not just an actuarial exercise. It was, as Bell emphasized, "evidence of the condition and prospects of the masses; and as such it affords a measure of their progressive improvement; or of their deterioration." Reformers thus urged the creation of state registration systems at least partly to ensure having such evidence.[11] But they also collected information about the environment in which the poor lived that reduced their expectancy, and then charted what needed to be changed. Accumulation of such statistics was assumed to play a direct and holy role in the "improvement of the moral and social condition of society."[12]

As in England, these kinds of data were disseminated more and more widely, not only to officials and physicians, but to the public. Middle-class Americans in the 1840s and 1850s could get such figures from reports, tracts, and public meetings, though some observers realized that the sanitary cause was not always well served by large amounts of statistics. At a Boston meeting in 1846, for instance, it was noticed that "the thread of statistical poverty was spun too fine and too long for the patience of a popular assembly, however necessary minute details may be in a pamphlet."[13]

Gouverneur Emerson, who had had long experience with the problems of the poor, reminded physicians that little sanitary progress could be expected "until the subject has been agitated again and again." And from his knowledge of the medical profession, he was doubtful if physicians as a group had the moral fervor or civic activism needed to do this. Essential for an effective reform movement, he felt, were totally dedicated "philanthropists," perhaps someone like Dorothea Dix, to seize upon the statistics of urban squalor and sickness and "bring them under the notice of the legislator. Where individuals have been led forward by a laudable zeal for the cause of humanity, it is truly astonishing how much can be accomplished by their means."[14]

Emerson did not really expect to find reformers of the same caliber or effectiveness as Dix in his generation of American physicians. Actually, urban public-health problems were of such a magnitude and complexity that the fact-finding process for each city would require scores of energetic individuals. Anyone who became involved with the process would have to confront not merely bad conditions and public apathy but the strenuous opposition of economically interested landlords. Physicians had little motivation for engaging in such struggles; they rarely thought of themselves as crusaders. Even those with the strongest humanitarian or social convictions had busy practices, institutional commitments, civic responsibilities, or scientific activities which kept them from engaging in the prolonged fact-finding and political agitation required in reform movements. And yet, within these limitations, a few doctors tried to get these processes started in their communities.

The Urban Sanitary Survey

A number of information-gathering projects conspicuously tried to emulate the comprehensive sanitary surveys of the British. Most of the projects were begun by the medical profession, though other groups also played important roles in some communities. In almost every case, however, American surveys fell far short of the British in scope, thoroughness, and amount of information unearthed.

Urban sanitary surveys were not really the same thing as the familiar medical-topographical survey, although they shared some aspects in common. Medical-topographical surveys had generally been concerned with laying bare the various elements of the *natural* environment that might have some effect on disease and mortality. The sanitary surveys, on the other hand, were aimed predominantly at revealing disease-causing conditions in the *artificial* environment of the city. These conditions included the poor air, poor water, and impure soil in and around the houses, streets, vacant lots, and places of work that man had created.

The earliest clear call for American urban sanitary surveys based on the English model came in New York City. In 1842 John Griscom was appointed City Inspector of New York, a post to which he brought experience as a dispensary physician and attending physician at New York Hospital, as well as a special interest in physiology and hygiene. His single year of work as inspector pointed up for him the need for better vital statistics and heightened his awareness of the poverty and sanitary problems in New York.

As City Inspector, Griscom issued a single annual report, which he later elaborated upon in various public lectures and in pamphlets. His best-known publication had a distinctly Chadwickian title, *The Sanitary Condition of the Laboring Population of New York,* but it was not, in any substantive sense, a systematic sanitary investigation like the ones Chadwick and his colleagues had made among the poor of Great Britain.[15] Rather, it was essentially a polemic, a plea for a comparably thorough survey to be made in New York. Griscom's visits to some of the incredibly foul cellars, where many poor people were forced to live, and other "pestiferous places" were the chief sources of his "exposé." He buttressed his impressions with extracts from Chadwick's report on the plight of the poor in Great Britain and from Horace Mann's call for the teaching of hygiene to the people. He also summarized the statistics of the medical care provided to the poor at the various New York hospitals and dispensaries and related them to the bills of mortality. And, he quoted replies to questionnaires that he had sent out to the most knowledgeable witnesses of New York's poverty: the dispensary physicians and the missionaries of the City Tract Society, energetic and dedicated men who moved freely among the most wretched.

Griscom's tract constituted a persuasive argument for effective sanitary policing of New York. It was obvious to him that "when individuals of the pauper class are ill, their entire support, and perchance that of the whole family, fall upon the community." New York's institutions in about 1850 were estimated to have some fifty-three thousand poor people, out of a population of about five hundred thousand, receiving free public medical care.[16] Improving the living conditions of these people would, among other things, both reduce their total sickness as a class and reduce the financial burden on the rest of the population.[17]

Griscom's exposé of New York's conditions, and his recommendations for official sanitary and housing inspection, were generally endorsed by the city's editors, physicians, and philanthropists.[18] They also stimulated other individuals and organizations to look further into some of the conditions. Of special interest was the 1850 police census of inhabited cellars. From it middle-class readers were shocked to learn that 18,456 individuals lived underground, in what the *New York Tribune* called a "subterranean city from whose damp and filthy portals ooze up the foul and poisonous miasma which continually pollutes the air, and sows the seeds of disease broadcast among the inhabitants of the Upper City." These people were the very worst off of a population that was still growing rapidly; the federal census of 1850 showed that the five hundred thousand persons were crammed into some thirty-seven thousand habitations, about 13.5 people per dwelling. To a generation for whom single-family houses were still the norm, this was "un-natural huddling [which] necessarily engenders disease, and demoralizes our population."[19]

Despite this and other inquiries, the fact remained that neither Griscom nor anyone else was able to organize a comprehensive sanitary survey in the city before the Civil War, let alone bring about an effective reform of its corrupt and archaic public-health provisions. Griscom was, however, able to help advance the idea of the urban sanitary survey in other areas of the United States through his participation in the reform efforts of the national medical community.

During the mid-1840s, ambitious members of the Medical Department of Washington's National Institute for the Promotion of Science tried to stimulate medical-topographical surveys, but they had relatively few responses to their questionnaires. Accordingly, when the American Medical Association was formed, the institute transferred most of its fact-finding efforts to the new body. The institute's delegates to the Baltimore meeting of the AMA in 1848, James Wynne and John M. Thomas, took with them not only a resolution supporting registration systems but one to establish a permanent committee to gather information on public hygiene. When it was established, this committee proved to be more interested in urban sanitary surveys than in medical-topographical studies.

The AMA's Committee on Public Hygiene, as originally constituted, included a number of physicians who had already demonstrated an interest in sanitary affairs: Wynne, Griscom, Edward H. Barton and E. D. Fenner of New Orleans, Josiah Curtis of Lowell, Isaac Parish of Philadelphia, and others. To fulfill its mission, the committee quickly drew up a brief list of questions aimed at illuminating the sanitary condition in particular cities. And, to set an example, the members proceeded to prepare replies on conditions in their own communities.

The published reports of these inquiries varied widely in length, scope, and amount of concrete information. Some, including those of Curtis for Boston and Lowell, and of Barton and Fenner for New Orleans, were relatively thorough and included a fair amount of statistics. Others were brief, impressionistic comments. However, taken together, they exerted considerable influence within the medical community, judging from the committees on public hygiene which sprang up in state and local medical societies and from the large numbers of sanitary reports which appeared during the 1850s in American medical journals.[20]

However numerous these sanitary inquiries of the 1840s and 1850s were, they were only small beginnings, the limited efforts of individuals or, at most, of local medical societies. Some inquiries, as those in New Orleans, failed miserably to bring about much permanent sanitary improvement. No inquiry was really comprehensive, and none really dealt with areas larger than an individual city or an occasional county. The problem, it was generally agreed, was the almost complete failure of the federal and state governments to get involved in health-related activities. This failure was a great source of frustration for physicians and statists in a hurry to emulate their English and European colleagues. The reasons for such neglect were difficult to understand; certainly there was no consistent policy of laissez faire in the states.[21] On the contrary, American state governments were deeply involved in a variety of surveys and information-gathering enterprises.

Jacksonian America was a veritable anthill of investigatory activity, much of it governmentally supported. Army explorers and mappers provided bases for new territorial claims. Railroad surveyors probed valleys and mountain passes to determine routes for new lines. Botanists and zoologists roamed ever further in their searches for specimens. Geological surveys, launched by a few states as early as the midtwenties, became common in other places. Surveys of agricultural resources were also numerous. Some states developed comprehensive programs to obtain information about various areas of knowledge. In 1836, for example, the New York legislature provided funds for a detailed survey and study of the state's natural history, a project that in scarcely a decade produced large scholarly reports on the state's zoology, mineralogy, geology, agriculture, and botany.

Shattuck had argued the need for comprehensive statewide public-health data throughout his campaign for the Massachusetts registration system, and he reemphasized the matter in his Boston census report of 1845. The lack of such information about the state was shocking, he felt, when he noticed how much had been done to map the physical resources of Massachusetts. "We have had a trigonometrical survey . . . a geological survey . . . an agricultural survey . . . a zoological survey . . . [but] where is the sanatory map which points out the healthy and unhealthy localities in the State."[22]

During the midforties, Shattuck and Jarvis explored the idea of a sanitary survey with members of the American Statistical Association and the Massachusetts Medical Society. By 1848 and 1849, respectively, these bodies had approved and sent memorials on the subject to the Massachusetts legislature. In May 1849 the legislature commissioned three investigators and authorized an expenditure of $500 for a "plan for a Sanitary Survey of the State." Shattuck himself was on the legislative committee that drew up the bill, and he was subsequently selected as one of the three commissioners, an appointment he later claimed was "entirely unsolicited and unexpected." Since the other two members of the commission, Nathaniel P. Banks and Jehiel Abbott, had no special qualifications for the project, Shattuck asked Jarvis to collaborate in devising the survey plan and writing the report. Jarvis ultimately declined, though he did provide advice at several stages of the work.

As a result, the document which went to the legislature in the spring of 1850 was in all essential aspects Shattuck's work.[23]

Shattuck's lengthy report was undoubtedly both less and more than his fellow legislators had bargained for. On the one hand it did not suggest any easy ways of financing local water systems, sewers, garbage removal, or other sanitary works, though it did urge Massachusetts to adopt British standards in constructing and operating such works. On the other hand, the report was not a mere protocol for a single sanitary inventory of Massachusetts communities, similar to those prepared for the AMA Committee on Public Hygiene, for by this time, with Great Britain's public health act of 1848 as an immediate example, Shattuck was thinking of something much larger and more useful. His recommendations thus emerged as a comprehensive plan for permanent state and local health offices and for public health activities guided by exhaustive provisions for continued data-collecting. Under the plan, periodic surveys of the overall sanitary condition of each separate locality would be provided for, but more significant would be the constant fact-finding involved in activities such as the inspection of housing, factories, and schools, nuisance abatement, vaccination, quarantine, supervision of foods, and the control of infectious diseases.

The reception of Shattuck's report was mixed. Physicians, and not only those in Massachusetts, reviewed the report in glowing terms and hoped that its suggestions could be implemented in every state.[24] Almost all of the provisions were ultimately accepted by subsequent generations as normal activities of health departments. For American lawmakers of Shattuck's generation, however, the proposals were just too much to accept. The Massachusetts legislature received the report in 1850 and ordered copies printed, but it failed to enact a law based on Shattuck's program. Indeed, not until well after the Civil War, a full decade after Shattuck's death in 1859, did the state adopt a similar public health plan.[25]

A more modest report than the one Shattuck presented to the state legislature in 1850, one that did not propose such a sweeping extension of state authority and that was closer to the original intent of a fact-finding survey, might have had a better political reception. A realistic and acceptable bill would have been one incorporating the more limited Baconian aspects of the plan, the provisions for a comprehensive state-run network of data gathering, record keeping, and education. One of Shattuck's central aims had been to create a system "designed to pile up fact upon fact, in relation to life, disease, and mortality, until their nature and laws are ascertained and demonstrated." Shattuck thus thought that the registration of vital statistics and the conduct of censuses should logically be brought together under the board of health. At the same time, the state and local boards would undertake a wide range of sanitary inquiries, meteorological observations, epidemiological studies, and other fact-finding activities. And, along with these, Shattuck envisioned that the state would disseminate health information on the broadest scale, hoping thereby to reduce the public burden of the sick poor and, by improving the "vital force" of the individual citizen, to fulfill the antebellum ideal of the best possible personal hygiene.

The failure of Massachusetts legislators to implement any part of Shattuck's report left the state without means to conduct thorough and regular sanitary surveys. In this respect Massachusetts was no different from other American states before the Civil War. As Jarvis confessed to Chadwick in 1853, "we have no light to offer you from America. We have no reports of sanitary surveys, so that we have no means of determining the sanitary condition of any place or people here."[26]

Urban Waste and Waste Removal

Despite this lack of formal reports, Americans could usually get a general idea of the conditions that prevailed in particular cities and of what was being done to improve them. They could readily quantify the wagonloads of garbage, stable manure, night soil, and trash which went daily to the dumps, or the amounts of water carted or piped in to clean the streets and help wash the household slops away. It was not even difficult to calculate how much open area existed between the habitations in a city, or how much living space per capita in one tenement. The important point was that almost no city did enough waste removal, especially in the areas inhabited by immigrants and other poor people. Likewise, little was done either to ensure adequate space between buildings or to restrict the numbers of inhabitants in a given building.

From the earliest public hygiene reports submitted to the American Medical Association one could learn that Concord, New Hampshire, like many other cities, had few if any paved streets. Only a few places, including New York, had paved virtually all of their streets and alleys. Some cities saw to it that their streets were cleaned at least occasionally, but others, including Portland, Maine, left them to be "washed by every shower." Larger places generally made contractual or other arrangements for cleaning the streets and removing refuse, but nowhere did they work well: Philadelphians complained that their alleys and unpaved streets got little attention, and in many cities, garbage and refuse piled up in the streets for the hogs and was only infrequently carted away, every three weeks in Cincinnati.

At this time, most American cities and towns had only surface gutters for the drainage of water and other wastes. Individuals and neighborhoods in a good many cities had their own private drains which emptied into rivers or ponds. A few cities had been constructing municipal sewers in some sections—by 1848 Boston had about 25 miles of sewers, Philadelphia nearly 12 miles, and Baltimore 1 mile. Here was a situation, the AMA's Committee on Public Hygiene agreed, where good statistics could undoubtedly provide an immediate stimulus to the extension of public sanitary works: "The bills of mortality of these different places, have not heretofore been kept in such a manner as to enable comparisons to be made between sewered districts and those which are without them. Could such results be obtained, the committee are well assured that they would present such facts, in relation to the high mortality of non-sewered districts, as would startle the legislators of our cities from their slumbers, and induce the rapid extension of these works to every portion of their respective towns."[27]

There was a comparable disparity between cities in their provisions for water. Most communities (58 percent of the 136 that had pipe systems in 1860) still relied on private companies to build and run their waterworks. Among the large cities, however, New York, Boston and Philadelphia had constructed extensive public systems, a trend that continued to grow. Across America, communities were engaging former canal or railroad engineers to make surveys for reservoirs, aqueducts, and pumping systems. Spokesmen, meanwhile, projected future water needs in terms of such factors as anticipated population and the numbers of hydrants to be required for fire-fighting. Because of constant immigration, these estimates could not be very accurate.

Some within the ranks of sanitary reform, notably Shattuck, advised cities, in the interest of fiscal responsibility, to go slow in extending their water systems.[28] But this was not the pre-vailing view among statists. The AMA Committee on Public Hygiene, for instance, was convinced "that many of the evils, and much of the disease incident to poverty, may be relieved by copious and never-failing supplies of pure water."[29] Walter Channing, Dean of the Harvard Medical School, similarly felt that water should be made available "even in wasteful pro-fusion." In 1843, Channing concluded that the Boston poor were filthy and intemperate most of the time because the water they had to purchase by the bucket was so expensive. "How can the poor acquire or preserve habits of cleanliness under such absolute destitution of its means? How strong is the temp-tation to *intemperance*, when the only sure means of *temperance* are so poorly supplied?"[30]

Temperance advocates were among the most vocal advo-cates of public-run water systems, for, though inadequate, they furnished more water at far less cost to the consumer than private systems did, frequently with free water outlets for the poor. When New York and Boston celebrated the inaugura-tion of their public systems in the 1840s, the temperance marching groups were as conspicuous in the parades as the militia, firemen, or politicians.

Physicians generally agreed that the new availability of water would be a great stimulus both to temperance and to health.[31] How great a stimulus it actually was did not seem measureable at first, but by the midforties, some figures had become avail-able on the relation of water to health. Data compiled at that time in Philadelphia showed that, following the improvement of the water supply, deaths from cholera infantum began to decline, apparently due to increased cleanliness of the inhabitants and of their homes.[32]

Generally speaking, most of the antebellum American correlations between dirt and mortality, and sanitary works and healthy towns were still only inferences drawn from the British statistics. This lack of data may have been a disadvantage for some sanitary reformers in arguing for new sanitary works or the enlargement of old ones, but in the large towns, the combined threat posed by immigration, industrialism, and imminent cholera was so great that the need for sanitary works was felt to be all too evident, even without precise statistical measurement of the problem. In any case, reformers could not wait for perfect public statistics, but made do with information that they rounded up themselves, either as individuals or in professional societies. With few exceptions, cities also had to limp along with makeshift public health offices and with part-time health officers to apply what statistics there were.

The Public Health Professional—Edwin M. Snow

If even Massachusetts, with its exceptionally enlightened physicians, statisticians, and lay reformers, could not enact any sort of comprehensive and workable health legislation before the Civil War, it was certainly beyond the capacity of the other states. Similarly, the demonstrated indifference of New York City's governing officials to sanitary problems was equally observable in most other cities. Small communities typically had few standing mechanisms, boards, or officials; their ordinances permitted town officers to take public health action but did not require it. Larger cities like Boston, New York, and Philadelphia had a variety of boards, medical advisory committees, and part-time doctors performing particular sanitary tasks and collecting some information, but these were rarely effectively coordinated. Few of the boards had consistent political support, and none had the full-time technically skilled personnel to do the job. More than many other reforms, sanitary reform demanded expertise and constant technical attention. Griscom and Shattuck thought that professional health officers with training in both sanitation and statistics were necessary if the abundant local laws and the policies of the boards of health were to be carried out properly.

Griscom felt that city health officers should be physicians, but with a background broad enough to be able to combine the medical care provided by the dispensary physician, the cleaning and preventive activities of the sanitarian, and the fact-finding of the surveyor or the statistician. The field duties of such an official would routinely "carry him in the very track of the nuisances which should be corrected," and would also make him a true health missionary as he took the rules of hygiene to the poor in their homes. But, Griscom thought, the health officer should also be a scientist charged with studying the distribution and causes of disease and death, keeping meteorological observations, and analyzing vital statistics. As a professional, he would be expected to make the local statistical reports useful for something "more than to satisfy ordinary curiosity, or to enable the preacher "to point a moral."[33]

Shattuck agreed with Griscom that the health officers of individual cities and towns should be well-educated physicians. They should have no allegiance to any particular theory of disease or therapy, should be devoted to the ideal of prevention, and should be "especially acquainted with sanitary science." They would govern their day-to-day work by information generated from annual sanitary surveys and from mortality data supplied by the local clerk or registrar.

In Shattuck's view, the state boards of health should not be composed of medical men entirely, because they would also need the expertise of lawyers, engineers, chemists, and other public men and professionals. Moreover, he could see no need for the principal state health officer to be a physician, since the officer would have no direct responsibilities with sick people. Still, this individual, like Horace Mann on the state Board of Education, would be the board's most knowledgeable expert, the full-time sanitary activator, administrator, and coordinator of the board's various programs. The qualifications for such a professional, Shattuck suggested, were dual: familiarity with public health problems and the etiology and prevention of disease, together with a thorough grounding in statistics, a skill essential to the duties of supervising registration, carrying out sanitary surveys, conducting censuses, preparing abstracts and reports, and disseminating sanitary information. Through such activities this office would necessarily become the most important statistical bureau in the state.

Shattuck felt that the extensive use of statistics in public health work was little more than an extension of methods found essential at home, in business, and in other areas of government. He defined statistics as "the science or art of applying facts to the elucidation and demonstration of truth." He himself professed to have "no particular fondness for figures," yet he was totally dedicated to the facts which they represented. For, he concluded, whatever distaste statistics might inspire, "in this 'matter-of-fact-age' they are required, and they are far more useful and important than the fiction and theory, the assumption and assertion, that have occupied so much of public attention."[34]

The first American community in which a combination of statistical, medical, and sanitary expertise was given a real chance to succeed in public health administration was Providence. The path to public health reform there was opened in 1852, when the Rhode Island state legislature, after several failures, enacted a bill which made workable provisions for the registration of vital statistics. This state registration system, which required the cooperation of the secretary of state and a committee of the state medical society, was perhaps not ideal, but, it rapidly proved to be one of the most effective in the United States. As long as the committee had competent and interested members, the arrangement produced a reliable supply of medically accurate data.[35]

Once in operation, registration soon proved remarkably useful in local sanitary reform. In fact, registration, in conjunction with the need for information on the cholera, prompted municipal officials of Providence to institute permanent sanitary mechanisms. During 1854, 159 deaths from cholera were registered in Providence. Edwin M. Snow, a dispensary physician, worked long hours among the poor during this outbreak. Early in the epidemic he persuaded the mayor to require physicians to file detailed information with the city clerk about the medical, social, and environmental circumstances of each cholera case. In the fall, following the epidemic, Snow collated these returns and studied them along with the mortality data furnished by the registration system. As a result, he was able to pinpoint 88 percent of the deaths as being among the foreign-born poor, and further, to associate roughly 70 percent of the deaths (111 out of 159) with two particular neighborhoods inhabited predominantly by foreigners. For the confirmed anticontagionist, as Snow was, two likely sources of the excess mortality of these neighborhoods suggested themselves immediately: in one area, the foul state of a canal; in the other, the "offensive effluvia from the hog-pens." In other parts of the city, deaths from cholera were attributed to a variety of filthy conditions in or near the habitations where the deaths occurred: offensive privies, foul gutters, dark and narrow lanes, malodorous drains, or the general absence of provisions for removing filth.

In January 1855 Snow summarized his statistics in a strong letter to the mayor of Providence. He argued that the city government was directly to blame for the mortality by having failed to collect and study the sanitary facts of the city and by neglecting to take appropriate preventive measures. To ensure that this would be done in the future, the city needed a knowledgeable health officer given appropriate authority, just as the city needed expertise in education, welfare, public works, and other concerns.[36]

Backed by an alarmed public and an unusually unified medical profession, Snow's proposal received quick attention from the mayor and city council, and by midyear it had been translated into new local sanitary ordinances. In May 1855 an act transferred the task of registering vital statistics from the overworked City Clerk's office to a new city registrar, who was now charged with proper study of the distribution and causes of death in the interest of public health. In July, Snow, who had become something of a local hero, was appointed to that office, as well as to the posts of quarantine officer and city health physician. By the end of the year, the city already felt a new sense of security, thanks to the vigor that Snow brought to these offices, and it was ready to give him anything he needed.[37] In December, to ensure continued public awareness of mortality and of specific sanitary evils, he sought and obtained authorization to publish regular registration reports for the city apart from those of the state. And in mid-1856 the city created for him the new position of Superintendent of Health.

Early in July 1856 Snow obtained a "nuisance complaint book" for the health office, to register unsanitary conditions needing investigation, and laid it on a table next to his registers of births, marriages, and deaths. With this symbolic act he stressed the fundamental underpinning of antebellum public health work in simple records and numbers. And as he carried out his duties he became the principal American example of a public health professional. In Great Britain, only Liverpool and London had installed professional medical officers of health before Snow was elected, but neither city entrusted the twin positions of registrar and health officer to the same individual.

In Providence, Snow did not become responsible for many actual sanitary operations, nor was he given legal authority to have these operations carried out. Neither he nor the community wanted the health professional to have such powers, but saw him instead as the city's resident scientific expert on what was to be done. As registrar, he carefully scrutinized every death return, got physicians to correct deficiencies, studied the accumulating data's significance for the health of Providence, and published his abstracts and analyses of data to ensure public awareness of the situation. As health officer, he investigated all kinds of filth, identified possible causes of disease, carried out public vaccinations, advised on the building and operation of municipal sanitary works and facilities, conducted systematic sanitary surveys of the city annually, and correlated his findings with the results of registration. Snow's annual published reports for the two offices indicated what progress had been made and still needed to be achieved.[38] And, as the various kinds of data accumulated, more and better generalizations became possible about the people of the community. A graph of the prospects of life in Providence, inspired by Farr's famous biometer and published by Snow in 1856, provides an example of how the data were used (see Figure 3).

With the successive steps taken in Providence following the cholera of 1854, observers remarked that the ratio of the city's deaths to population went into a noticeable decline: from 1 in 36.6 in 1854, to 1 in 46.1 in 1856, and 1 in 55.7 in 1857. At the same time, editors and statists quickly came to recognize the vital statistics of Providence as among the most thorough

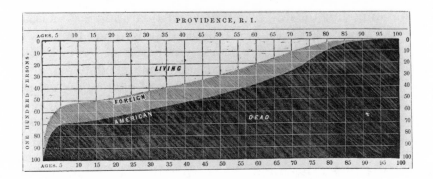

Figure 3. An American "Biometer." The Providence Superintendent of Health, Dr. Edwin M. Snow, explained this diagram as follows: "This diagram represents the progress through life of a generation of persons born in Providence, of American, and of another of foreign parentage, at the present rate of mortality of the two classes. The space of lighter shade between the two lines marked "American" and "foreign," represents the excess of mortality of the foreign over the American population. The vertical lines represent the different ages from birth to 100 years, and the point of intersection of any vertical line with the two lines marked "American" and "foreign," shows the number living and dead at that age, of each 100 persons born. Take, for example, the vertical line at 15 years; its intersection with the two curved lines shows that, at 15 years after birth, of each 100 persons of American parentage, 67 would be living and 33 would be dead; while of each 100 persons of foreign parentage, 50 would be living and 50 dead. At 40 years after birth, there would be 53 living and 47 dead of American parentage, and 36 living and 64 dead of foreign parentage in each 100; at 70 years, American 26 living, 74 dead; foreign 10 living, 90 dead . . . Similar diagrams illustrating the mortality in different parts of England may be found in the Fifth Annual Report of the Registrar General." Source: Edwin M. Snow, *Second Annual Report on the Births, Marriages and Deaths in the City of Providence, for the Year ending December 31, 1856* (Providence: Knowles, Anthony, 1857), p. 11. Courtesy of the National Library of Medicine.

and reliable in the United States. They agreed that almost overnight it had become "one of the best regulated cities in the country, in a sanitary point of view." Only Philadelphia in 1859 seemed to have a lower ratio of deaths to population. Only Boston had such effective sanitary police.[39]

Other American cities were still groping toward a systematic approach to their sanitary problems when the first national convention on quarantine was held in 1857 in Philadelphia. The physicians and city officials who attended, overwhelmingly holding anticontagionist views, felt that the time had come to simplify, ease, and make uniform the diverse restrictions impeding the flow of commerce in quarantined American ports. These men had little reliable data to guide them as to the relation of quarantine to various diseases. Nevertheless, during this and three subsequent conventions, they managed to reach some agreement on the practice of quarantine, though substantial improvement of the situation had to wait several more decades for federal action.[40] At the same time, the delegates were not long satisfied with restricting themselves to this one problem, "external hygiene." As a result, beginning with the 1858 convention, they turned their attention increasingly to sanitary matters, or the "internal hygiene" of cities. The conventions eventually created committees to work on a wide variety of sanitary problems. Among their accomplishments at the 1859 meeting, the delegates adopted Henry G. Clark's sanitary code as a model for state and local public-health legislation. And in 1860 they endorsed Snow's suggestions for uniform registration procedures.[41]

As active health officers, Clark and Snow occupied special roles of authority in the successive meetings. Snow, to his undoubted embarrassment, was singled out by one enthusiastic participant as "some extraordinary specimen of humanity, for what he has accomplished in the sanitary regulation" of his city.[42] There were indeed few others who could be called public health professionals. However, just by bringing a large number of interested individuals together, the convention made its contribution towards professionalism. Although the fact-finding processes of public health investigation were not yet very far developed, some health officers were already looking beyond death registration to the registration of sickness and disease.[43] Most, however, with John Bell, just wanted the vital statistics registers to become operational and the sanitary surveys systematized. They hoped for the day when sanitary reformers in the United States would no longer have to depend on foreign statistics for their ammunition.[44]

10

The Computing Mind at Midcentury

In their increasingly large requirements for data, public health workers were somewhat more representative of antebellum statistical enthusiasts than were the clinical physicians, who dealt with relatively limited amounts of facts and figures. Over-all, however, the proliferations of numerical applications in medicine closely resembled those in many other antebellum enterprises. Taken together, by 1860 these varied enterprises, as fact-finding agencies, were producing veritable torrents of data that needed sorting and analysis. Statistical activity had become an essential part of the day-by-day conduct of government and business; it lent authority to every social and educational reform movement; it was an indispensable tool for the sciences; and it was fast becoming an obsession in ordinary people's lives, a development which philosophers were beginning to perceive as a dubious blessing. Statistics seemed to have come of age.

The Ubiquitous Statistic

In 1860 Oliver Wendell Holmes perceived that the "observing and computing mind," in its rise during the century, had made the terms *law* and *average* into "the two dominant words of our time . . . Statistics have tabulated everything—population, [economic and industrial] growth, wealth, crime, disease. We have shaded maps showing the geographic distribution of larceny and suicide. Analysis and classification have been at work upon all tangible and visible objects."[1]

Medical analysts of the antebellum generation could feel satisfied that with the diffusion of statistics the Baconian processes of observation and induction had become indispensable to medical progress. In fact, many agreed with Joseph LeConte that statistics "was the only means of applying induction to a large portion of our science."[2] In any case, with the Baconian ideals now generally taken for granted on all sides, fewer antebellum Americans found it necessary to invoke Bacon for support in arguing for quantitative investigations. The accomplishments of a generation of living statistical standard-bearers—Quetelet and Babbage, Louis and Farr, Shattuck and Jarvis—now provided the authority needed.

Along with their European peers, American medical statisticians argued whether statistics was a science in its own right or merely an auxiliary to the various sciences. Most of them seemed to think of it as a separate science, and in any event, it was steadily taking on the characteristic appendages of professionalism, a process in which Americans became increasingly involved. They prepared analytical reviews of European statistical publications for their journals. They discussed new techniques in their societies. And, from 1853 onward, some of them began to meet with their European peers at the newly launched International Statistical Congresses. Jarvis, who visited London in 1860, dined at the Thatcher House Tavern with members of the London Statistical Society, inspected the General Register Office, and "talked shop" with such leaders of the British statistical community as William Farr, Edwin Chadwick, and Florence Nightingale. All this, together with his participation in that year's international congress, constituted what he characterized as "the crowning pleasure . . . of my life."[3]

By 1860, the numerical tables and charts which Jarvis and his fellow statisticians laboriously produced came to be widely thought of as veritable "lighthouses" for guiding American society, business, government, and science. And yet, as Holmes's Poet-Autocrat suggested, these new authorities, the statisticians themselves, had too much dusty dryness about them to be considered among the heroes of the age.[4] Moreover, not everyone was willing to believe that the universe was governed solely by the prosaic and deterministic laws that emerged from statistical thinking.

Romantics conspicuously opposed quantification because it tended to reduce the traits and capabilities of the individual to a mean, to a common denominator. Among them, Ralph Waldo Emerson was well aware that, through the examination of sufficiently large numbers, society could perceive the laws of population, wars, diseases, weather, and many another phenomenon. Moreover, he acknowledged that the state had a paramount obligation to apply these laws to secure "the greatest good of the greatest number" of its citizens. However, it was important, at least for poets and philosophers, to avoid the general and impersonal characteristics of "l'homme moyen" and in their place to cultivate individualistic modes of living. They could do this by the free exercise of intellect and feeling, by concerning themselves less with the material aspects of life and more with the intangible and spiritual elements. While the understanding and operation of man's world required statistical computation, the world of nature, Emerson asserted, "hates calculators; her methods are saltatory and impulsive . . . The results of [a truly intuitive] life are uncalculated and uncalculable." Neither religious belief nor philosophical insight was capable of being quantified or calculated. And certainly poetic creativity did not require the "arithmetic of statists."[5]

Limits of Midcentury Medical Statistics

Meanwhile, by 1860 certain physicians were also questioning whether even medical investigation should rely as heavily upon statistical method as it had come to in recent decades. Some were beginning to suggest that medicine could benefit from intuition and deduction from limited facts, as well as from Baconian induction from large numbers of observed facts. It was not a question of reverting to the well-remembered bad old days when simplistic and impressionistic medical systems held sway, for it was clear that statistical thinking and critical numerical methods had been essential elements in the impressive recent elevation of medicine and most of its branches— surgery, obstetrics, hygiene, public health, pathology, and the behavioral sciences, among others—far beyond their earlier state. Yet, it had to be admitted that statistical analysis had not yet done much to solve some of medicine's most fundamental problems, including working out the etiology of diseases and providing authoritative assessments of the various forms of therapy. Other modes of investigation were clearly needed, not to displace statistics totally but to supplement it, at least in some areas of medicine.

By midcentury it was realized that publications which made heavy use of statistics were not necessarily either informative or useful. Critics were beginning to suggest, moreover, that the process of induction did not require unusual mental ability and, indeed, that statistical analysts were not always the brilliant thinkers they had been made out to be. Some bluntly concluded, in fact, that much of the statistics then produced was "the work of laborious and slow minds."[6] In short, the early easy acceptance of statistical or numerical method began to be replaced by an increasingly critical attitude toward its technical shortcomings and its newly dominant overall role in orthodox thinking.

Among the sources of dissatisfaction with statistics was the tendency of writers to reduce medical matters tediously to numerical components when the application of ordinary common sense would have been quite sufficient. Many important medical matters were being neglected while excessive attention was being bestowed upon counting trivia. This practice was sometimes traced to the need of editors to fill up the pages of their new medical journals. By the forties, physicians in every section of the country found themselves progressively pressed by editors to jot down something about their local medical experiences. Whatever their training or ability, if they could write and count they were virtually assured of getting into print. Snatching a few moments from busy practices, such individuals hastily tabulated the results of some phase of their practices over a few years, committed them to paper, and sent them off for publication, no doubt often hoping that publication would enhance their careers. These articles were unfortunately often extremely crude and superficial.[7]

By contrast with this deliberate stimulation of statistics, some editors adopted a policy of not including meaty statistical materials in their journals. As late as 1852 Southern editors as a group were reportedly reluctant to include tables or rows of figures because they feared that readers would find them unappealing. Furthermore, Southern printers were reputed to be so prone to error that any statistical data published in Southern journals was virtually useless.[8]

At the same time, thinking people were increasingly skeptical of medical numbers because of the kinds of wild claims and obvious misuses that had led Luther Bell and Isaac Ray to denounce the statistics of insane asylums. Even the defenders of statistics sometimes found it difficult to support a method that was so consistently misapplied, though, as Bennet Dowler saw it at midcentury, statistics was the medical profession's best substitute for chaos and mere conjecture: "If numerical results be as yet unsatisfactory, if but few probabilities and still fewer truths have been eliminated, and if some errors have been propagated in the name of numerism, the fault is not in the method, but in its misapplication."[9]

While America's medical inquiries sometimes failed to observe even the simple numerical standards set by Louis, they were even more deficient in terms of the standards of the emerging statistical science. Some physicians were aware of the professional evolution of statistics—of the growing statistical literature, of the development of increasingly refined analytical techniques, of the activities of statistical societies, or even, from the early fifties, of the international congresses of statisticians—but their ranks were few. Some might appreciate, but few could construct, the sort of equations with which William Farr increasingly filled his analyses of the Registrar General's data on disease. Few grounded their studies firmly on the authority of significantly large quantities of numbers. Few figured or studied the mean values of their numbers. All too many were content with piling observation upon observation, and in the process, all too often they gave small attention to the analysis and interpretation of the accumulated facts.[10]

True, the medical profession as a whole did not yet worry very much about statistical niceties. Some of its spokesmen, however, were seriously concerned about the legitimacy of the all-embracing claims that were being made for numerical methods in medical science. Critics began to question those who were touting statistics as "a sort of machine, which is to set all things right, by the use of which error and mistake become impossible, the necessity for intellectual exertion is removed, and no occasion is offered for the exercise of the higher faculties of our nature."[11] They complained that the numerical enthusiasts tended to be one-sided and "extravagant" in their conclusions, that they neglected other ways of approaching medical problems.[12]

Bartlett's comprehensive vision of medical numeration was particularly open to criticism. Some faulted it as a distortion of the true Baconian induction, and others saw it as claiming too much for numerical method, as constructing a medical philosophy around number alone.[13] In fact, by the fifties even Jacob Bigelow and other early advocates were freely admitting that they had gone overboard in their earlier enthusiasm for medical statistics.[14]

A central criticism raised against Bartlett and his fellow proponents of numerical method was their assumption that hypothesis could play no part in scientific medicine. Henry Jacob Bigelow, a son of Jacob, summarized the point in an address to Harvard medical students in 1846. He argued that the application of inductive method should not be limited to numerical summaries of large amounts of facts; it was a process that every thinker might well use informally in testing some hypothesis. Bigelow conceded that any suggestion at this time of resurrecting the medical use of *hypothesis,* a word "suggestive of unsoundness and instability in science, [would] startle the severer disciple of the statistic school." Nevertheless, he pointed out, it could scarcely have been possible even for a Louis, the father of numerical method, to have influenced clinical medicine without having had some hypothesis in mind prior to or along with his collecting and analysis of medical observations. The trouble was that Louis's followers tended to be unaware of this or denied that it was true. According to Henry Bigelow, too many of them had adopted the form of numerical method without pursuing the substance, without applying it to legitimate medical *ideas,* including speculations. Lacking such intellectual purpose, he thought, the energetic collecting of medical statistics was all too frequently a sterile exercise.[15]

The more enthusiastic partisans of numerical method predictably maintained that Henry Bigelow's position was decidedly "open to objection." Stillé, for one, vehemently protested a viewpoint that was so fundamentally "at variance with that of Dr. Bartlett." It seemed to Stillé, in fact, that the younger Bigelow was getting perilously close to committing medical treason by "throwing open the door to that arch-enemy of medical science, Hypothesis."[16]

Actually, there was little danger that hypothesis could get far in the strongly empirical framework of orthodox medicine just before the Civil War. Spokesmen for rational medicine did begin to acknowledge that hypotheses leading to possible alternative investigative techniques could conceivably be helpful accompaniments to statistics. Yet, in the next breath they reminded everyone of the disastrous period before Louis, when medical hypothesis had had its day and had turned out to be not only useless but misleading and retrograde. In the nineteenth century only numerical method had contributed much to medicine; eschewing hypothesis, it had become the most productive source "of our knowledge of the laws of disease, of the events which make up their individuality, of their diagnostic characters, and, to a greater or less extent, of the effects of remedies."[17]

Nevertheless, despite its past contributions, and despite its continuing central importance to physicians, statistics had undoubtedly reached a plateau in its applications to clinical and investigative medicine. Its capacity to expand its areas of contribution to medicine was temporarily circumscribed. To some extent, the limitations were inherent in the very youth of the science. More conspicuously, they also derived from the mathematical shortcomings of antebellum physicians and statists.[18] And unfortunately, there seemed to be little chance of increasing mathematical knowledge significantly in the near future. American educational institutions then offered relatively few opportunities to excel in the subject. One 1852 essayist found that, except for the United States Military Academy, American colleges then afforded "little or no instruction in mathematics," and the same was probably equally true of medical schools.[19]

The truth of it was that most Americans saw little or no need to improve their command of mathematics beyond an elementary level. Up to this time, even engineers, business-men, and mechanics could usually get by with only rough calculations performed according to rules of thumb, and sim-ilarly, reformers, statists, and physicians needed only a mod-icum of arithmetic to produce the simple kinds of calculations that characterized most nineteenth-century statistics.

Partly because these elementary calculations left as many unanswered questions as they solved, midcentury medical sci-entists increasingly turned to other forms of investigation. In particular, they began to be attracted to problems that could be investigated in laboratories using the new developments in chemistry and physiology, and new equipment, such as the achromatic microscope. Few then thought of abandoning the one method of investigation for the other, but rather of using both as the occasion presented. In New Orleans, Bennet Dow-ler, for example, thus undertook a series of numerical studies of febrile caloricity that he intended as models for examining other physiological phenomena, but he also followed with keen interest the reports of European laboratory work in organic chemistry, cellular biology, and other fields which were ex-panding the horizons of biomedicine. Likewise, John William Draper enthusiastically drew upon Quetelet's statistics of human growth for use in his 1856 physiology text, while he also in-troduced his own photomicrographs to demonstrate bodily functions.[20] And Joseph Jones, perceiving that the enterprise of medical research was moving beyond the competence of single investigators, urged that statisticians of his day join with microscopists, chemists, and other specialists in scientific teams which could examine malaria and other diseases carefully and from many viewpoints.[21]

Well before 1850 even Austin Flint was keeping a close watch on the findings which were emerging from European laboratories, particularly those in physiology. In 1854, con-scious of the limitations of his own numerical studies, he ar-ranged to spend several months in Paris to study other modes of investigation. There he paid special attention to the exper-imental medicine of Claude Bernard. Bernard had come to the conclusion that it was futile to expect much from statistics, at least as long as its methods remained so inadequately developed. For the time being, he urged medical scientists to focus their efforts upon qualitative rather than quantitative studies.[22]

After the Civil War, this viewpoint spread rapidly through the American medical community, and by the last decades of the nineteenth century, the laboratory effectively dominated medical research. Statistical analysis receded, temporarily but decidedly, into the background. It was not until the twentieth century that investigators began finally to gain the competence in higher mathematics that was needed for statistics once again to play a major role in clinical and scientific studies.

NOTES

INDEX

Notes

Primary and secondary sources that directly support this study are cited in the notes. Scholars may obtain from me, upon request, a classified bibliography of secondary sources I consulted in my research.

1. America's Early Health Inventory

1. For colonial data-gathering activities, see James H. Cassedy, *Demography in Early America: Beginnings of the Statistical Mind, 1600–1800* (Cambridge, Mass.: Harvard University Press, 1969). For a study of emerging statistical consciousness about 1800, see Patricia Cline Cohen, "Statistics and the State: Changing Social Thought and the Emergence of a Quantitative Mentality in America, 1790 to 1820," *William and Mary Quarterly,* 3d ser., 38 (1981), 35–55.

2. E. S. Pearson, ed., *The History of Statistics in the 17th and 18th Centuries, against the Changing Background of Intellectual, Scientific, and Religious Thought. Lectures by Karl Pearson given at University College London during the Academic Sessions 1921–1933* (New York: Macmillan, 1978), pp. 2–9.

3. Timothy Dwight, *Travels in New England and New York*, ed. Barbara Miller Solomon, 4 vols. (Cambridge, Mass.: Harvard University Press, 1969), I, 58.

4. Samuel Williams, *The Natural and Civil History of Vermont* (Walpole, N.H.: Thomas and Carlisle, 1794), p. 328.

5. Dwight, *Travels,* I, 59–60.

6. Ibid., III, 293.

7. Benjamin Henry Latrobe, *The Journal of Latrobe* (New York: D. Appleton, 1905), pp. 33–34, 131–132.

8. Benjamin Rush, "An Inquiry into the Comparative State of Medicine, in Philadelphia, between the Years 1760 and 1766, and

the Year 1809," in his *Medical Inquiries and Observations,* 4th ed., 4 vols. in 2 (Philadelphia: M. Carey, 1815), IV, 235–236.

9. Ibid., IV, 235–239.

10. For a contemporary account of the epidemic in one city, see William Currie, *A Sketch of the Rise and Progress of the Yellow Fever, and of the Proceedings of the Board of Health, in Philadelphia, in the Year 1799* (Philadelphia: Budd & Bartram, 1800).

11. See, for instance, City of Baltimore, *A Series of Letters and other Documents relating to the Late Epidemic of Yellow Fever* (Baltimore: William Warner, 1820); and Currie, *Sketch of the Yellow Fever.*

12. Valentine Seaman, "Inquiry into the Cause of the Prevalence of the Yellow Fever in New York," *Medical Repository,* 1 (1797 / 98), 315–332. For subsequent uses of spot maps as adjuncts of statistics in plotting yellow fever and, to a lesser extent, cholera, see Lloyd G. Stevenson, "Putting Disease on the Map: The Early Use of Spot Maps in the Study of Yellow Fever," *Journal of the History of Medicine,* 20 (1965), 226–261.

13. R. LaRoche, *Yellow Fever,* 2 vols. (Philadelphia: Blanchard & Lea, 1855).

14. Benjamin Waterhouse, *A Prospect of Exterminating the Small-Pox . . .* (Cambridge: William Hilliard, 1800); Benjamin Waterhouse, *A Prospect of Exterminating the Small Pox, Part II* (Cambridge, Mass.: William Hilliard, 1802).

15. Valentine Seaman, *A Discourse upon Vaccination, or Kine-Pock Inoculation* (New York: Samuel Wood & Sons, 1816), p. 50. See also Seaman's "Cow-Pock Inoculation in New York," *Medical Repository,* 10 (1807), 430–431; John Redman Coxe, *Practical Observations on Vaccination: or Inoculation for the Cow-Pock* (Philadelphia: James Humphreys, 1802).

16. C. F. Volney, *A View of the Soil and Climate of the United States,* trans. & ed. Charles Brockden Brown (Philadelphia: J. Conrad, 1804), pp. 223–264; David Warden, *A Statistical, Political, and Historical Account of the United States of North America,* 3 vols. (Edinburgh: Archibald Constable, 1819), I, 256–283.

17. Anthony Fothergill, "Meteorological Observations on the Climate of Philadelphia, with General Remarks on Population, Longevity, Diseases, &c.," observations kept between 1806 and 1810. Unpub. MS at Library of College of Physicians, Philadelphia.

18. Unsigned review, *New England Journal of Medicine and Surgery,* 1(1812), 180.

19. "Some Account of the Disease that was Epidemic in some Parts of New York and New England, in the Winter of 1812–13," *New England Journal of Medicine and Surgery*, 2 (1813), 241–243; *London Medical Review*, 5 (1812), 67; "Superior Salubrity of the North-American Climate, *Medical Repository*, 6 (1803), 432; Dwight, *Travels*, IV, 217–224; and W. P., "Health Statistics," *Niles' Register*, 8 (1815), 257.

20. Frederick Dalcho, "Anniversary Oration," *Philadelphia Medical Museum*, 1 (1804 / 1805), 125–141; David Ramsay, *The Charleston Medical Register for the Year MDCCCII* (Charleston: W. P. Young, 1803); Joseph Johnston, *An Oration Delivered before the Medical Society of South Carolina, at their Anniversary Meeting, Dec. 24th, 1807* (Charleston: Marchand, Willington, 1808).

21. David R. Arnell, "A Geological and Topographical History of Orange County, New York," *Medical Repository*, 12 (1809), 313–323; Nicholas Romayne and John Stearns, "Circular Letter of the Medical Society of the State of New York to the County Societies," *Medical Repository*, 12 (1809), 396–397; and "New York Prize Medals," *Medical Repository*, 11 (1808), 327. The society hoped to work with recently formed agricultural, natural history, and other societies to coordinate health-related inquiries. In other states, some of the newer learned societies, including the Connecticut Academy of Arts and Sciences and The Columbian Institute, proposed to prepare comprehensive topographical and statistical histories, but their members proved unable to go very far with them.

22. [Samuel Latham Mitchill et al.], "Circular Address," *Medical Repository*, 1 (1797 / 98), vii–xii. David Ramsay expressed a similar outlook in launching the *Charleston Medical Register;* see vol. 1 (1803), pp. iii-iv.

23. David Warden, "Observations on the Natural History of the Village of Kinderhook, and its Vicinity," *Medical Repository*, 6 (1803), 4–18.

24. Samuel Latham Mitchill, *The Picture of New York* (New York: Riley, 1807); James Mease, *The Picture of Philadelphia* (Philadelphia: Kite, 1811); Daniel Drake, *Notices Concerning Cincinnati* (Cincinnati: John W. Browne, 1810); Daniel Drake, *Natural and Statistical View, or Picture of Cincinnati and the Miami Country* (Cincinnati: Looker and Wallace, 1815).

25. The first three censuses provided six types of population information: heads of families, free white males aged sixteen and over, free white males under sixteen, free white females, all other free persons (except untaxed Indians), slaves. In subsequent censuses,

information was gathered on aliens (1820); deaf, "dumb" (mute), and blind (1830); insane and "idiots" (1840); and mortality (1850).

26. The first bill issued by the parish appeared in 1737. Subsequent bills appeared almost every year from 1748 through 1774, and then again after the Revolution. Similar bills were published during the 1750s by Saint Philips Church in Charlestown, South Carolina.

27. For an early appreciation of these reports, see Gouverneur Emerson, "Medical Statistics," *American Journal of the Medical Sciences,* 1 (1827), 116–117.

28. Cassedy, *Demography in Early America,* pp. 237–241, 251–255.

29. City of Boston, *Rules, Regulations and Orders of the Board of Health of Boston . . .* (Boston: Board of Health, 1814), sections 1 and 7, pp. 7–8.

30. For a detailed discussion of the emergence of Boston's provisions for vital statistics see Lemuel Shattuck, "On the Vital Statistics of Boston," *American Journal of the Medical Sciences,* n.s. 1 (1841), 369–401; see also John B. Blake, *Public Health in the Town of Boston, 1630–1822* (Cambridge, Mass.: Harvard University Press, 1959), pp. 213–215.

31. See State of New York, *Health Laws of New York* (New York: James Cheetham, 1805), pp. 30–31. New York City Board of Health, *Extracts from the Various Laws Relative to the Preservation of Health in the City of New York* (New York: Southwick & Pelsue, 1811), pp. 3–4.

32. Noah Webster, "Number of Deaths, in the Episcopal Church in New York, in each month for ten years—from January 1, 1786, to Dec. 31, 1795," *Memoirs,* Connecticut Academy of Arts and Sciences, vol. I, pt. I, (1810), pp. 97–98.

33. John Pintard, "Bill of Mortality for New York," *Medical Repository,* 6 (1803), 97–98.

34. Samuel Latham Mitchill, "Summary View of the Modes by which Human Life terminates in the City of New York," *Medical Repository,* 11 (1808), 35.

35. Samuel Latham Mitchill, "Abstracts from the Bills of Mortality kept in the City of New York, during 1807 and 1808," *Medical Repository,* 13 (1810), 335.

36. John Linnaeus Edward Whitridge Shecut, *Flora Carolinaeensis,* 2 vols. (Charleston: Hoff, 1806); Benjamin Smith Barton, *Collections for an Essay towards a Materia Medica of the United States* (Philadelphia: Way & Groff, 1798).

37. Erasmus Darwin, *Zoonomia; or The Laws of Organic Life,* 2 vols.

(London: Johnson, 1794–1796); Nathaniel Chapman, *Discourses on the Elements of Therapeutics and Materia Medica,* 2 vols. (Philadelphia: Webster, 1817), pp. 60 ff.

38. *The Pharmacopoeia of the United States of America, 1820* (Boston: Wells & Lilly, 1820).

39. Pinel, for one, made nosology an integral part of his larger efforts to achieve medical precision. As a tool for organizing the diseases he encountered in the hospital, it became a frame of reference for analyzing and evaluating the therapies and clinical regimens he used there.

40. Among the individuals who spoke out were Pierre Jean Georges Cabanis in France and Thomas Jefferson in the United States.

41. John Brown, *Elementa medicinae* (Edinburgh: Elliot, 1780), p. 421. There were at least eight American editions of Brown's work by 1806.

42. Unsigned review, *Medical Repository,* 17 (1815), 42; Benjamin Rush, *Medical Inquiries and Observations,* 4th ed., III, 19–20.

43. Edward Miller, "An Attempt to deduce a Nomenclature of certain Febrile and Pestilential Diseases from the Origin and Nature of their Remote Cause," *Medical Repository,* 7 (1804), 362–372; John Beale Davidge, *Nosologia Methodica* (Baltimore: Edes, 1812); David Hosack, *A System of Practical Nosology* (New York: Van Winkle, 1818); Franklin Bache, "Review of Several Nosological Arrangements," MS dated February 1, 1823, Bache Papers, College of Physicians, Philadelphia.

44. Davidge, *Nosologia,* p. xliv.

45. Hosack, *Nosology,* pp. vii–viii.

46. Daniel Drake, "Practical Essays on Medical Education and the Medical Profession in the United States," *Western Medical and Physical Journal,* 3 (1829 / 30), 333; and William Currie, *A Synopsis, or General View of the Principal Theories or Doctrines of Diseases* (Philadelphia: Parker, 1815), pp. 162–163.

47. William Currie, *A View of the Diseases Most Prevalent in the United States of America, at Various Seasons of the Year* (Philadelphia: Humphreys, 1811).

48. Spalding, *Reflections on Fevers,* p. 4; Rush, *Medical Inquiries,* 4th ed., III, 23 ff.

49. Horatio Gates Jameson, *The American Domestick Medicine, or Medical Admonisher* (Baltimore: Lucas, 1817), p. 116; Samuel Latham Mitchill, "Summary View," p. 33.

50. [John D. Godman], "Nosology," *Western Quarterly Reporter of Medical, Surgical and Natural Science,* 1 (1822), 111–122; unsigned

review, *New England Journal of Medicine and Surgery*, 8 (1819), 72–75.

2. A Calculus of Medical Philanthropy

1. "M. N.," quoted in *The Panoplist*, 14 (1818), 153.

2. Timothy Dwight and Lyman Beecher made their own surveys. Timothy Dwight, *Travels in New England and New York*, ed. Barbara Miller Solomon, 4 vols. (Cambridge, Mass.: Harvard University Press, 1969), IV, 322–326. Beecher's survey was summarized in David B. Warden, *A Statistical, Political, and Historical Account of the United States of North America*, 3 vols. (Edinburgh: Constable, 1819), III, 479–487. "Summary View of Protestant Missions," *Missionary Herald*, 24 (1828), 1–15, 33–41.

3. *Missionary Herald*, 23 (1827), 139.

4. *The Autobiography of Benjamin Rush*, ed. George W. Corner (Philadelphia: American Philosophical Society, 1948), p. 243.

5. Samuel Latham Mitchill, "Abstracts from the Bills of Mortality kept in the City of New York, during 1807 and 1808," *Medical Repository*, 13 (1810), 336; "Condition of the Poor House in the City of New-York, 1812–13," *Medical Repository*, 16 (1813), 315.

6. Alexander H. Everett, *New Ideas on Population*, 2d ed. (Boston: Cummings, Hilliard, 1826), pp. 95, 105–108.

7. James Fenimore Cooper, *Notions of the Americans*, 2d ed., 2 vols., (New York: Stringer & Townsend, 1850), I, 98.

8. "Pauperism and Gratuitous Attendance," *Medical Magazine*, 3 (1834), 521.

9. J. Roby, "Boston Dispensary," *Medical Magazine*, 3 (1834), 438.

10. "Means by which Diseases from hard Drinking and Venereal Virus are Promoted," *Medical Repository*, 7 (1804), 90–91. See also the massive 1859 report, based on a census of prostitutes and many interviews, of William W. Sanger, *The History of Prostitution* (New York: Harper, 1859).

11. Barbara Gutmann Rosenkrantz, "Booby Hatch or Booby-Trap: A New Look at Nineteenth-Century Reform," *Social Research*, 20 (1972), 733–743.

12. "Intemperance and Disease," *Boston Medical and Surgical Journal*, 40 (1836), 263.

13. *A Brief Account of the New-York Hospital* (New York: Collins, 1804), pp. 5–6.

14. Ibid., pp. 64–65; "New York Hospital Returns for 1802–3," *Medical Repository*, 7 (1804), 293. "Information concerning the New-

York Hospital," *Medical Repository*, 14 (1811), 189; and extracts from hospital reports, *North American Medical and Surgical Journal*, 10 (1830), 439, 442. By 1836 it was estimated that "more than 30,000 patients are annually provided for at the public expense in this city [New York]; of these at least nine-tenths are foreigners." Charles A. Lee, "Medical Statistics," *American Journal of the Medical Sciences*, 19 (1836 / 37), 27. See also "Atmospheric Constitution and Diseases of New-York, April 30, 1818," *Medical Repository*, 19 (1818), 299.

15. Charles Caldwell, "Thoughts on the probable destiny of New Orleans, in Relation to Health, Population, and Commerce," *Philadelphia Journal of the Medical and Physical Sciences*, 6 (1823), 1–14.

16. Thomas W. Griffith, "History of the Baltimore Almshouse," written about 1826, in Douglas G. Carroll, Jr., and Blanche D. Coll, "The Baltimore Almshouse: An Early History," *Maryland Historical Magazine* (1971), pp. 135–152, but especially pp. 145 and 149.

17. John B. Brown, "Account of a Fever which prevailed in the Boston Almshouse in 1817–18," *New England Journal of Medicine and Surgery*, 7 (1818), 116.

18. J. H. Miller, quoted in an editorial, "Report of Dr. J. H. Miller," *Thomsonian Recorder*, 4 (1835 / 36), 151. "An Account of Patients ill of Malignant Fever, received at Bellevue Hospital, near New-York, in the Summer and Autumn of 1803," *Medical Repository*, 8 (1805), 147.

19. Early clinical research in some of the hospitals and almshouses will be examined in the next two chapters.

20. For an early, flawed estimate, see J. M. Toner, "Statistics of Regular Medical Associations and Hospitals of the United States," *Transactions of the American Medical Association*, 24 (1873), 314–333. During the 1830s, Samuel Jackson of Philadelphia thought that: "Our greatest evil lies in the paucity of our hospitals, and the miserable management, as it regards the great interests of science, under which they are placed." Samuel Jackson, *The Principles of Medicine* (Philadelphia: Carey & Lea, 1832), p. xvii.

21. See "Pauperism, Hospitals," *Buffalo Medical Journal*, 5 (1849 / 50), 303–309. The author of this editorial, probably Austin Flint, pointed out, however, that most American communities were then spending so much on their almshouses that they could not afford to build hospitals as well.

22. In a variation of this, the founders of the Boston Lying-In Hospital carefully avoided giving any encouragement to illegitimacy by "limiting in the strictest manner, its benefits to married women." "Boston Lying-In Hospital," *Medical Magazine*, 2 (1833), 19.

23. "Pennsylvania Hospital," *North American Medical and Surgical Journal*, 10 (1830), 439; "Massachusetts General Hospital," *Boston Medical and Surgical Journal*, 16 (1837), 112; ibid., 24 (1841), 51; Matthew Clarkson and Thomas Buckley, *State of the New-York Hospital for the Year 1813* (New York, 1814), p. 3.

24. Samuel Cartwright, communication in *Report of the Select Committee of the Senate of the United States on the Sickness and Mortality on Board Emigrant Ships* (Washington: Tucker, 1854), p. 135.

25. The growth of hospitals for the insane will be dealt with in Chapter 7.

26. *Thirteenth Report of the Directors of the American Asylum at Hartford, for the Education and Instruction of the Deaf and Dumb*, May 16, 1829 (Hartford: Hudson and Skinner, 1829), pp. 12–13; *Address of the Trustees of the New-England Institution for the Education of the Blind* (Boston: Carter, Hendee, 1833), pp. 4–5.

27. Thomas H. Gallaudet, *A Discourse* (Hartford: Hudson & Co., 1821). In 1819 the institution's name was changed to the American Asylum at Hartford, for the Education and Instruction of the Deaf and Dumb.

28. Samuel Akerly, *Address Delivered . . . 30th May, 1826* (New York: Conrad, 1826), pp. 18–20.

29. The proportion of blind reported in the 1830 census was slightly greater than this, or 1 in 2,363. Of the 6,106 listed as deaf in this census, 743 were Negroes.

30. Lewis Weld, *An Address delivered . . . February 16th, 1828* (Washington: Way & Gideon, 1828), passim.

31. Beck noted, from the state census reports, an increase in New York of from 644 in 1825 to 1,070 in 1835, or from a proportion of 1 deaf-mute for every 2,510 persons to 1 in 2,032. T. Romeyn Beck, "Statistics of the Deaf and Dumb in the State of New-York, the United States, and in various countries of Europe," *Transactions of the Medical Society of the State of New York*, 3 (1836 / 37), 330, 334.

32. *Thirteenth Report of . . . the American Asylum at Hartford*, pp. 13–14.

33. *Address of the Trustees of the New-England Institution for the Education of the Blind to the Public* (Boston: Carter, Hendee, 1833), p. 3–4.

34. Samuel Hazard, "Statistical Observations on the number of blind in Pennsylvania and the United States," *Western Journal of the Medical and Physical Sciences*, 12 (1838), 44–58.

35. [Daniel Drake?], unsigned review, *Western Journal of the Medical and Physical Sciences*, 12 (1838), 88–89.

36. "The Massachusetts Charitable Eye and Ear Infirmary," *Medical Magazine*, 1 (1833), 154–157.

37. Beecher felt that intemperance had an adverse effect upon health and physical energy, intellect, military prowess, patriotism, moral principle, and industry and economy, in addition to leading to irreligion. Lyman Beecher, *Six Sermons on Intemperance* (Boston: Marvin, 1828), pp. 48–56.

38. Gouverneur Emerson, "An Account of an Epidemic Fever, which prevailed among the Negroes of Philadelphia, in the Year 1821," *Philadelphia Journal of the Medical and Physical Sciences*, 3 (1821), 200.

39. Beecher, *Sermons on Intemperance*, pp. 54–55.

40. Benjamin Rush, "An Inquiry into the Comparative State of Medicine, in Philadelphia, between the Years 1760 and 1766, and the Year 1809," in his *Medical Inquiries and Observations*, 4th ed., 4 vols. in 2 (Philadelphia: M. Carey, 1815), IV, 236.

41. "Means by which Diseases from Hard Drinking and Venereal Virus are Promoted," *Medical Repository*, 7 (1804), 89.

42. "Preface," *Medical Repository*, 19 (1818), iii–iv.

43. [Richard H. Coolidge, ed.], *Statistical Report on the Sickness and Mortality in the Army of the United States, 1855-1860* (Washington: Bowman, 1860), p. 97.

44. William P. C. Barton, *Hints for Naval Officers Cruising in the West Indies* (Philadelphia: Littell, 1830), pp. 3–51.

45. William M. Wood, "Practical Reflections on the Grog Ration," in his *Shoulder to the Wheel of Progress* (Buffalo, 1863), pp. 138–139.

46. D [Daniel Drake], "Progress of Temperance," *Western Journal of Medicine and Surgery*, 7 (1843), p. 79.

47. [Samuel Forry, comp.], *Statistical Report on the Sickness and Mortality in the Army of the United States . . . January, 1819, to January, 1839* (Washington: Gideon, 1840), p. 218; Samuel Forry, *The Climate of the United States and its Endemic Influences*, 2d ed. (New York: Langley, 1842), pp. 326–328.

48. David Ramsay, *The History of South Carolina*, 2 vols. (Charleston: Longworth, 1809), II, 418; J. E. White, "A Few Remarks on the Weather and Diseases of 1805," *Medical Repository*, 11 (1808), 23. "Report of the Committee appointed to investigate the causes and extent of the late extraordinary sickness and mortality in the town of Mobile," *Medical Repository*, 20 (1820), 340.

49. Forry, *Climate of the United States*, pp. 326–327.

50. Benjamin Rush, *An Inquiry into the Effects of Spirituous Liquors*

on the Human Body (Boston: Thomas & Andrews, 1790), p. 12.

51. D. Humphreys Storer, "Medical Statistics and Bills of Mortality, for Boston, during nineteen years ending January 1st, 1832," *Medical Magazine*, 1 (1833), 516; Charles A. Lee, "Medical Statistics," pp. 34–35. Gouverneur Emerson noted similar underreporting or concealment of certain causes of death in Philadelphia during this period.

52. Beecher, *Sermons on Intemperance*, pp. 48–49, 97–98.

53. Benjamin Rush, *Medical Inquiries and Observations upon the Diseases of the Mind* (Philadelphia: Kimber & Richardson, 1812), p. 33; see also a review in *Western Journal of Medicine and Surgery*, 2d ser., 7 (1847), 180.

54. "Report of the Committee appointed to investigate the causes and extent of the late extraordinary sickness and mortality in the town of Mobile," *Medical Repository*, 20 (1820), 340.

55. Unsigned review, *Western Journal of Medicine and Surgery*, 2d ser., 7 (1847), 180–181; Martyn Paine, "History of the cholera at Montreal," *Boston Medical and Surgical Journal*, 8 (1833), 53–65.

56. "Record of Cholera Cases," *Boston Medical and Surgical Journal*, 7 (1832), 237–238; James W. Stone, "The Cholera Epidemic in Boston," *Boston Medical and Surgical Journal*, 41 (1849), 298–299.

57. D [Daniel Drake], "Progress of the Temperance Reform," *Western Journal of Medicine and Surgery*, 2 (1840), 318–319. See also "Physiological Temperance Society," *Boston Medical and Surgical Journal*, 25 (1842), 114. About ten years prior to this a Medical Student's Temperance Society was formed at the University of Pennsylvania.

58. J, "Intemperance and Disease," *Boston Medical and Surgical Journal*, 15 (1836), 261.

59. Ibid., pp. 261–267.

60. John Bell, *On Regimen and Longevity* (Philadelphia: Haswell & Johnson, 1842), pp. 374–376; D, "Progress of the Temperance Reform," pp. 318–319; S. C. Farrar, "General Report on the Topography . . . of Jackson . . . Mississippi," *Southern Medical Reports*, 1 (1849), 351; John C. Warren, *The Preservation of Health* (Boston: Ticknor, Reed, & Fields, 1854), pp. 117–118.

61. Samuel A. Cartwright, "Hygienics of Temperance," *Boston Medical and Surgical Journal*, 48 (1853), 373–377. From his own medical records, Cartwright published similar quantitative analyses on the extent of intemperance among the lawyers and the general populace of Natchez; see pp. 494–499; and ibid., 49 (1853), 9–13. In summary, out of 790 of his adult male patients in Natchez, he classified 330 as temperate and 460 as intemperate. "As the women are

all temperate," he added, "no notice was taken of them in collecting the statistics" (p. 9).

62. D, "Travelling Editorials," *Western Journal of Medicine and Surgery*, 8 (1843), 237–238.

63. Quoted in Bell, *On Regimen and Longevity*, pp. 365–368.

3. Quantification in Orthodox Medicine

1. Gradually expanding versions of Rush's system appeared in the successive editions of his *Medical Inquiries and Observations*, starting with the first edition (vol. I, Philadelphia: Prichard & Hall, 1789; vol. II, Philadelphia: Dobson, 1793).

2. For details of this growth, see Glenn Sonnedecker, *Kremers' and Urdang's History of Pharmacy*, 4th ed., rev. (Philadelphia: Lippincott, 1976). The only other branch of American medicine that attained a comparable professional level before the Civil War was dentistry, which enjoyed also considerable international standing as early as the 1840s.

3. B. L. S., review, *Philadelphia Journal of the Medical and Physical Sciences*, 10 (1825), 162–163.

4. The extent of this approach in Great Britain has been examined in detail by U. Tröhler, "Quantification in British Medicine and Surgery, 1750–1830" (Ph.D. diss., University of London, 1978).

5. Samuel Miller, *A Brief Retrospect of the Eighteenth Century*, 2 vols. (New York: Swords, 1803), I, 202, 529.

6. Jefferson to Caspar Wistar, June 21, 1807, *The Writings of Thomas Jefferson*, ed. Paul Leicester Ford, 10 vols. (New York: Putnam, 1892–1899), X, 78–85; Pierre Jean Georges Cabanis, *Du degré de certitude de la médecine* (Paris: Didot, 1798).

7. "Circular Address," *Medical Repository*, 1 (1797 / 98), vii.

8. Gouverneur Emerson, "Medical Statistics," *American Journal of the Medical Sciences*, 1 (1827), 116–117; John D. Godman, *Contributions to Physiological and Pathological Anatomy* (Philadelphia: Carey, 1825).

9. John Bell, "Some general remarks on the use of the Stethoscope," *New-York Medical and Physical Journal*, 3 (1824), 269–281; Samuel G. Morton, *Illustrations of Pulmonary Consumption, its Anatomical Characters, Causes, Symptoms and Treatment* (Philadelphia: Key & Biddle, 1834).

10. See, for example, William L. Lytton, *An Inaugural Dissertation on Dropsy* (New York: Swords, 1807); Felix Pascalis, "Observations on the Ulcerated Tonsils of Children," *Medical Repository*, 13 (1810),

19–25; Horace H. Hayden, "Anatomical and Pathological Observations on the Teething of Infants, etc.," *Medical Repository*, 13 (1810), 217–225.

11. James Jackson, *Remarks on the Brunonian System* (Boston: Wait, 1809); George Fordyce, "An Attempt to improve the Evidence of Medicine," *Transactions of a Society for the Improvement of Medical and Chirurgical Knowledge* (London: Johnson, 1793), I, 243–293. Jackson acknowledged Fordyce's example in 1836. See his Preface to Pierre-Charles-Alexandre Louis, *Researches on the Effects of Bloodletting in some Inflammatory Diseases,* ed. James Jackson (Boston: Hilliard, Gray, 1836).

12. J. Augustine Smith, "A Lecture on Medical Philosophy," *New York Medical and Physical Journal,* 7 (1828), 174–183.

13. Samuel Jackson, "Clinical Reports of Cases treated in the Infirmary of the Alms-House of the City and County of Philadelphia," *American Journal of the Medical Sciences,* 1 (1827), 85–109; and *An Introductory Lecture to the Institutes of Medicine* (Philadelphia: Sharpless, 1830), especially pp. 14–15, 26–35.

14. Russell Jones has found evidence that six hundred or more Americans studied in Paris between 1820 and 1861. Most had earned their medical degrees before going to Paris. Russell M. Jones, "American Doctors in Paris, 1820–1861: A Statistical Profile," *Journal of the History of Medicine and Allied Sciences,* 25 (1970), 143–157.

15. J. Barnes, review, *American Journal of the Medical Sciences,* 4 (1829), 404.

16. Henry I. Bowditch to his sister Mary, Nov. 1833, in Vincent Y. Bowditch, *Life and Correspondence of Henry Ingersoll Bowditch,* 2 vols. (Boston: Houghton Mifflin, 1902), I, 64.

17. Quoted in Henry I. Bowditch to his mother, Jan. 27, 1833, in Vincent Y. Bowditch, *Life of Bowditch,* I, 33.

18. See James Jackson, *A Memoir of James Jackson, Jr., M.D.* (Boston: Butts, 1835), pp. 20–24, 37–38; and Henry I. Bowditch to his Mother, Jan. 27, 1833, in Vincent Y. Bowditch, *Life of Bowditch,* I, 37–38.

19. G. P. [George Parker], letter to the editor, *Boston Medical and Surgical Journal,* 18 (1838), 74; Jackson, *Memoir of James Jackson, Jr.,* p. 17.

20. Similarly, Louis ultimately pointed to the Americans as being among the most productive and effective followers of his method, particularly in the study of typhoid fever. See his "Avertissement," in his *Recherches anatomiques, pathologiques, et thérapeutiques sur . . . fièvre typhoïde,* 2d ed., 2 vols. in 1 (Paris: Baillière, 1841), I, xvi–xvii.

21. Bowditch to his mother, Jan. 27, 1833, and to his father, Feb. 6, 1833, in Vincent Y. Bowditch, *Life of Bowditch*, I, 36–39, 62.

22. Jules Gavarret, *Principes généraux de statistique médicale* (Paris: Becket et Labé, 1840).

23. Erwin H. Ackerknecht, *Medicine at the Paris Hospital, 1794–1848* (Baltimore: The Johns Hopkins University Press, 1967), pp. 121–127.

24. Oliver Wendell Holmes, *The Position and Prospects of the Medical Student* (Boston: Putnam, 1844), p. 14.

25. William W. Gerhard, "Reports of Cases Treated in the Philadelphia Hospital at Blockley," *American Journal of the Medical Sciences*, 18 (1836), 301–325; "Remarks on Clinical Instruction," ibid., 387–392; and *Clinical Guide and Syllabus of a Course of Lectures on Clinical Medicine and Pathology* (Philadelphia: Auner, 1837), pp. 4–7.

26. Victor J. Fourgeaud, "Eclecticism in Medicine," *Saint Louis Medical and Surgical Journal*, 3 (1845 / 46), 98–107; E. M. Pendleton, "Statistics of Disease in Hancock County," *Southern Medical and Surgical Journal*, n.s. 5 (1849), 393; Theophilus Mack, "A Summary of the Claims of the Medical Art to the Consideration of the Public at the Present Day," *Buffalo Medical Journal*, 14 (1858 / 59), 456–458.

27. James Jackson, Sr., Preface to Louis, *Researches on the Effects of Bloodletting*, p. ix.

28. Editorial, attributed to John C. Warren, "New Theory of Deformity," *Boston Medical and Surgical Journal*, 3 (1830), 642–643.

29. Jules Gavarret, *Principes généraux*.

30. Elisha Bartlett, *An Essay on the Philosophy of Medical Science* (Philadelphia: Lea & Blanchard, 1844).

31. J. C. N. [Josiah C. Nott], review, *New Orleans Medical Journal*, 1 (1844 / 45), 490–494; A. Stillé, review, *American Journal of the Medical Sciences*, n.s. 16 (1848), 398–406.

32. Jacob Bigelow, "Brief Exposition of Rational Medicine," in his *Modern Inquiries: Classical, Professional, and Miscellaneous* (Boston: Little, Brown, 1867), pp. 244 ff. See also Alfred Stillé to George Cheyne Shattuck, May 7, 1842, Shattuck Papers, vol. 17, Massachusetts Historical Society.

4. Numerical Method and Rational Medicine

1. Jacob Bigelow, "Brief Exposition of Rational Medicine," in his *Modern Inquiries: Classical, Professional, and Miscellaneous* (Boston: Little, Brown, 1867); Worthington Hooker, *Rational Therapeutics* (Bos-

ton: Wilson, 1857); and Austin Flint, "Conservative Medicine as Applied to Therapeutics," *American Journal of the Medical Sciences*, n.s. 45 (1863), 22–43.

2. Worthington Hooker, *Physician and Patient* (New York: Baker & Scribner, 1849), p. 219.

3. W. H. [Worthington Hooker], review, *American Journal of the Medical Sciences*, n.s. 37 (1859), 505; and review, ibid., 40 (1860), 473.

4. Unsigned review, *American Journal of the Medical Sciences*, n.s. 16 (1848), 367–368; Alfred Stillé, *Elements of General Pathology* (Philadelphia: Lindsay & Blakiston, 1848), pp. 33–48; E. H. C. [Edward H. Clarke], review, *American Journal of the Medical Sciences*, n.s. 40 (1860), 131.

5. Samuel Forry, "Statistics of Re-vaccination," *American Journal of the Medical Sciences*, n.s. 3 (1842), 365–372; Caspar W. Pennock, "Note on the Frequency of the Pulse, and Respiration of the Aged," *American Journal of the Medical Sciences*, n.s. 14 (1847), 68–75.

6. Louis to James Jackson, Jr., Oct. 28, 1832, in James Jackson, Sr., *A Memoir of James Jackson, Jr., M.D.* (Boston: Butts, 1835), pp. 20–22.

7. Ibid., pp. 55–57.

8. William W. Gerhard, "On the Typhus Fever, which occurred at Philadelphia in the spring and summer of 1836," *American Journal of the Medical Sciences*, 19 (1837), 289–322. Gerhard acknowledged that H. C. Lombard of Geneva had asserted the distinction a few months earlier, but noted that Lombard had not provided a persuasive body of evidence to support his finding.

9. James Jackson, Sr., *A Report founded on the Cases of Typhoid Fever, or the Common Continued Fever of New-England, which Occurred in the Massachusetts General Hospital, from the Opening of that Institution, in September, 1821, to the End of 1835* (Boston: Whipple & Damrell, 1838); Enoch Hale, *Observations on the Typhoid Fever of New England* (Boston: Whipple and Damrell, 1839); Elisha Bartlett, *The History, Diagnosis, and Treatment of Typhoid and of Typhus Fever* (Philadelphia: Lea & Blanchard, 1842).

10. Samuel G. Morton, *Illustrations of Pulmonary Consumption, its Anatomical Characters, Causes, Symptoms and Treatment* (Philadelphia: Key & Biddle, 1834), pp. x–xi.

11. Pierre-Charles-Alexandre Louis, *Pathological Researches on Phthisis*, ed. Henry I. Bowditch (Boston: Hilliard, Gray, 1836), p. xxii.

12. "Outline of Remarks by A. Flint, M.D., of Buffalo, N.Y.," *Boston Medical and Surgical Journal*, 24 (1841), 231–233; Austin Flint, "Clinical Report on Dysentery; based on an analysis of forty-nine recorded cases," *Buffalo Medical Journal*, 9 (1853 / 54), 109–111.

13. Austin Flint, "Observations on the Pathological Relations of the Medulla Spinalis," *Buffalo Medical Journal*, n.s. 7 (1844), 269–271; Austin Flint, "On Variations of Pitch in Percussion and Respiratory Sounds, and their Application to Physical Diagnosis," *Transactions of the American Medical Association*, 5 (1852), 75–123.

14. Austin Flint, "Account of an Epidemic Fever which occurred at North Boston, Erie County, N. Y., during the months of October and November, 1843," *American Journal of the Medical Sciences*, n.s. 10 (1845), 21–35.

15. Jacob Bigelow, "On Self-Limited Diseases," in his *Modern Inquiries*, pp. 164–169.

16. Jackson, Sr., *Report on Typhoid Fever*, p. 23.

17. Pierre-Charles-Alexandre Louis, *Researches on the Effects of Bloodletting in some Inflammatory Diseases*, ed. James Jackson (Boston: Hilliard, Gray, 1836), pp. v–vii, and p. 171. "Dr. Jackson's Appendix to Louis on Bloodletting," *Boston Medical and Surgical Journal*, 14 (1836), 17.

18. [A. Flint?], "Remarks on Blood-letting," *Buffalo Medical Journal*, 2 (1846 / 47), 377–388, 450–455.

19. Bloomingdale Asylum for the Insane, *Extracts from the 24th Annual Report* (New York: 1844), p. 7; Pliny Earle, "Bloodletting in Mental Disorders," *American Journal of Insanity*, 10 (1853 / 54), 287–405.

20. D. J. Cain, "On the Use of Chinoidine in Periodical Fever," *Charleston Medical and Surgical Journal*, 15 (1860), 441–446; J. Chester Morris, "Experiments made to determine the Protective Power of Belladonna in Scarlatina," *American Journal of the Medical Sciences*, n.s. 33 (1857), 334–336; James B. Colegrove, "Oedema in Intermittent Fever," *Boston Medical and Surgical Journal*, 53 (1855 / 56), 114–117. Colegrove does not state what the overall proportion of inebriates in the almshouse was.

21. "Synopsis of Methods of Treating Asiatic Cholera," *Western Journal of Medicine and Surgery*, 3d ser., 3 (1849), 337–339; unsigned review, *New York Journal of Medicine*, n.s. 5 (1850), 241–243; W. L. Sutton and Jacob Bigelow, "On Large Doses of Calomel in Cholera, Etc.," *Boston Medical and Surgical Journal*, 41 (1849 / 50), 15–17.

22. A. G. Henry, "Epidemic Malignant Typhus," *Boston Medical and Surgical Journal*, 41 (1849 / 50), 15. "Cod-Liver Oil in Consumption," ibid., 50 (1854), 365. William P. Hort, "Congestive Fever," *New Orleans Medical and Surgical Journal*, 4 (1847 / 48), 58–60.

23. Austin Flint, "On the Treatment of Intermitting Fever," *American Journal of the Medical Sciences*, n.s. 2 (1841), 277–292; and "On the Employment of Quinia in Large Doses," *New York Journal of Medicine*, 6 (1846), 202–208.

24. "Extracts from an Official Report of J. J. B. Wright . . . on the Use of Large Doses of Sulphate of Quinine in Diseases of the South," *New York Journal of Medicine*, 5 (1845), 168–175. "Reports on the Administration of Quinine in Large Doses," *Statistical Report on the Sickness and Mortality in the Army of the United States . . . from January, 1839, to January, 1855* [ed. Richard H. Coolidge], (Washington: Nicholson, 1856), pp. 637–690.

25. Bennet Dowler, review, *New Orleans Medical and Surgical Journal*, 14 (1857), 426–427.

26. "Statistics of Cholera," *Buffalo Medical Journal*, 12 (1856 / 57), 317; Jackson, *Report on Typhoid Fever*, pp. 1–2.

27. "Facts and Fallacies," *American Medical Times*, 1 (1860), 443.

28. Alfred Stillé, *Elements of General Pathology*, pp. 33–46; Samuel D. Gross, review, *Western Journal of the Medical and Physical Sciences*, 12 (1838), 70–71.

29. Austin Flint, "Cases treated at the Erie County Almshouse," *Boston Medical and Surgical Journal*, 29 (1843), 390; and "Report of Clinical Observations on Continued (Typhus and Typhoid) Fever," *Buffalo Medical Journal*, 6 (1850 / 51), 131–134.

30. John Prosser Tabb, "Statistics of the Causes of Death in the Philadelphia Hospital, Blockley, during a period of twelve years," *American Journal of the Medical Sciences*, n.s. 8 (1844), 363, 374.

s75 31. Austin Flint, letters to the editor, *Buffalo Medical Journal*, 10 (1854), 109, 196, 329–330.

32. Daniel Drake, *An Introductory Lecture on the Means of Promoting the Intellectual Improvement of the Students and Physicians, of the Valley of the Mississippi* (Louisville: Prentice & Weissinger, 1844), passim; and "Western Medical Schools," *Western Journal of the Medical and Physical Sciences*, 9 (1835 / 36), 607–611. Henry D. Shapiro and Zane L. Miller, eds., *Physician to the West: Selected Writings of Daniel Drake on Science and Society* (Lexington: University Press of Kentucky, 1970), passim.

33. "Physicians' Account Book," *Boston Medical and Surgical Journal*, 55 (1856 / 57), 46–47; D. F. C. [Condie], review, *American Journal*

of the Medical Sciences, n.s. 32 (1856), 466–467; unsigned review, *Buffalo Medical Journal,* 11 (1855 / 56), 506–507; "Facts and Fallacies," *American Medical Times,* 1 (1860), 443; Henry Hartshorne, *Medical Record for Private Medical Statistics* (Philadelphia: Price, 1859).

34. John George Metcalf, "Statistics in Midwifery," *American Journal of the Medical Sciences,* n.s. 6 (1843), 328–329.

35. Ibid., pp. 327 ff. George N. Burwell, "Statistics and Cases of Midwifery; compiled from the Records of the Philadelphia Hospital, Blockley," ibid., n.s. 7 (1844), 317–319. "Obstetrical Statistics," *Buffalo Medical Journal,* 6 (1850 / 51), 433–434. William Ingalls, "Record of Obstetrical Cases," *Boston Medical and Surgical Journal,* 58 (1858), 233–236.

36. W. H. Gantt, "Table and Remarks on Private Obstetrical Practice as compared with Hospital Practice," *Saint Louis Medical and Surgical Journal,* 12 (1854), 216–221.

37. Henry A. Ramsay, "Contributions to Obstetrics," *Medical Examiner,* n.s. 6 (1850), 561–566. See also "Unreliable Obstetrical Statistics," *Boston Medical and Surgical Journal,* 45 (1850 / 51), 47; "The American Medical Association—Drs. Ramsay and Robertson," *Western Journal of Medicine and Surgery,* 3d ser., 8 (1851), 365–367; N. S. Davis, *History of the American Medical Association* (Philadelphia: Lippincott, Gambo, 1855), pp. 93–95. Ramsay reported at this time that white women of Georgia were beginning to turn away from the use of midwives.

38. John Swain, "Midwifery Statistics," *Western Journal of Medicine and Surgery,* 3d ser., 9 (1852), 401–402; 'Pacific' [James Blake?], "California Obstetrics," *Boston Medical and Surgical Journal,* 61 (1859 / 60), 169–170; James Blake, "Remarks on a Case of Rupture of the Vagina," *Pacific Medical and Surgical Journal,* 1 (1858), 141–145.

39. R. T. Trall, *The Hydropathic Encyclopedia,* 2 vols. in 1 (New York: Fowlers & Wells, 1854), II, 439. "Demonstrative Midwifery," *Buffalo Medical Journal,* 5 (1849 / 50), 565–566, 621–622; "Demonstrative Midwifery," *Western Journal of Medicine and Surgery,* 3d ser., 5 (1850), 459–460, 539–542; ibid., 7 (1851), 275–276.

40. J. P. Maynard, "Puerperal Convulsions—Expectant Treatment—Recovery," *Boston Medical and Surgical Journal,* 63 (1860 / 61), 369–374; F. Minot, "On Hemorrhage . . . in New-Born Infants," *American Journal of the Medical Sciences,* n.s. 24 (1852), 320; Chandler Gilman, "Report of the Committee on Obstetrics," *Transactions of the American Medical Association,* 2 (1849), 237–239; James D. Trask, "Statistics of Placenta Praevia," ibid., 8 (1855), 593–689.

41. Hugh L. Hodge, "Cases and Observations regarding Puer-

peral Fever," *American Journal of the Medical Sciences,* 12 (1833), 325–327; Charles D. Meigs, *On the Nature, Signs, and Treatment of Childbed Fevers* (Philadelphia: Blanchard & Lea, 1854), pp. 79–141.

42. Oliver Wendell Holmes, "The Contagiousness of Puerperal Fever," in his *Medical Essays, 1842–1882* (Boston: Houghton Mifflin, 1889), pp. 103–172.

43. Meigs, *Childbed Fevers,* passim; and *Woman: Her Diseases and Remedies,* 3d ed., rev. (Philadelphia: Blanchard & Lea, 1854), pp. 38–39, 591–608, 618.

44. Holmes, "Contagiousness of Puerperal Fever." The 1855 version of this paper was published separately under the name *Puerperal Fever as a Private Pestilence* (Boston: Ticknor & Fields, 1855). Most historians, too, have been thoroughly persuaded by Holmes's argument from probabilities. Recently, however, the pediatric historians Dorothy Lansing and Robert Penman in an unpublished paper have argued from modern information that puerperal fever is really a venereal infection (group 5 streptococcus) not spread by attending physicians.

45. J. C. W. [John Collins Warren], letter dated Dec. 30, 1837, *Boston Medical and Surgical Journal,* 18 (1838), 42–45.

46. Thomas F. Betton, "Statistics of Amputation," *American Journal of the Medical Sciences,* n.s. 11 (1846), 549.

47. George W. Norris, "Statistical account of the cases of Amputation performed at the Pennsylvania Hospital from January 1st, 1831, to January 1st, 1838," *American Journal of the Medical Sciences,* 22 (1838), 356–365; George W. Norris, "Statistical Account of the Cases of Amputation performed at the Pennsylvania Hospital from January 1, 1838 to Jan. 1, 1840," *American Journal of the Medical Sciences,* 26 (1840), 35–37.

48. Paul F. Eve, "Remarks on the Statistics of Amputation," *Southern Medical and Surgical Journal,* n.s. 2 (1846), 465–469.

49. Editorial, *Medical Examiner,* 1 (1838), 290–291; ibid., 2 (1839), 538; ibid., 3 (1840), 298–301.

50. Henry W. Buel, "Statistics of Amputations in the New York Hospital, from January 1st, 1839, to January 1st, 1848," *American Journal of the Medical Sciences,* n.s. 16 (1848), 33–43.

51. Frederick D. Lente, "A Statistical and critical account of the Fractures occurring in the New-York Hospital during the period of twelve years, elapsing between the 1st of January, 1839, and the 1st of April, 1851," *New York Journal of Medicine,* n.s. 7 (1851), 154–179.

52. Frank H. Hamilton, "Notes of an European Tour—No. 3,"

Buffalo Medical Journal, 1 (1845 / 46), 51–55.

53. Ibid., p. 55; "Fracture Tables," *Buffalo Medical Journal,* 4 (1848 / 49), 651–658; and "Deformities after Fractures," *Transactions of the American Medical Association,* 8 (1855), 349–354.

54. "Fracture Tables," *Boston Medical and Surgical Journal,* 49 (1853 / 54), 25; review, *Western Journal of Medicine and Surgery,* 3d ser., 12 (1853), 309–312.

55. Samuel D. Gross, *A Practical Treatise on the Diseases and Injuries of the Urinary Bladder, the Prostate Gland, and the Urethra* (Philadelphia: Blanchard & Lea, 1851). "Lithotomy and Calculous Diseases," *Western Journal of Medicine and Surgery,* 3d ser., 11 (1853), 334–335. Review, *New York Journal of Medicine,* n.s. 8 (1852), 105, 122–123. Samuel D. Gross, "Circular, to the Physicians of Kentucky," *Western Journal of Medicine and Surgery,* 4th ser., 3 (1855), 371–372.

56. George H. Lyman, *The History and Statistics of Ovariotomy* (Boston: Wilson, 1856). "Ovariotomy—its Statistics and Rate of Mortality," *Boston Medical and Surgical Journal,* 60 (1859), 280–283.

57. Jonathan Mason Warren, "Inhalation of Ether," *Boston Medical and Surgical Journal,* 36 (1847), 149–162; Walter Channing, *A Treatise on Etherization in Childbirth* (Boston: Ticknor, 1848); Paul H. Eve, "Report of Operations performed under Anesthetic Agents," *Southern Medical and Surgical Journal,* n.s. 5 (1849), 278–281; William T. G. Morton, "Comparative Value of Sulphuric Ether and Chloroform," *Boston Medical and Surgical Journal,* 43 (1850 / 51), 109–119. "Report of a Committee of the Boston Society for Medical Improvement, on the Alleged Dangers which accompany the Inhalation of the Vapor of Sulphuric Ether," ibid., 65 (1861 / 62), 229–242.

5. Varieties of Alternative Medicine

1. Thomas Hersey, *The Mid-wife's Practical Directory,* 2d ed. (Baltimore: Thomas Hersey, 1836), p. 48; John Swain, "Midwifery Statistics," *Western Journal of Medicine and Surgery,* 3d ser., 9 (1852), 401–402; unsigned review, *North American Medical and Surgical Journal,* 12 (1831), 171; Stephen Williams, "Climacteric Disease," *Boston Medical and Surgical Journal,* 52 (1855), 69–71; Robley Dunglison, *Elements of Hygiene* (Philadelphia: Lea & Blanchard, 1835), pp. 158–160.

2. "Law of Menstruation," *Boston Medical and Surgical Journal,* 14 (1836), 190–191.

3. Benjamin Rush, "An Inquiry into the Comparative State of Medicine, in Philadelphia, between the Years 1760 and 1766, and the Year 1809," in his *Medical Inquiries and Observations,* 4th ed., 4 vols. in 2 (Philadelphia: M. Carey, 1815), IV, 248.

4. [Samuel Latham Mitchill?], "Preface," *Medical Repository,* 19 (1818), iii.

5. James Fenimore Cooper, *The American Democrat,* in his *Representative Selections,* ed. Robert E. Spiller (New York: American Book, 1936), pp. 224–226. See also "Atmospheric Constitution, Diseases, and Bills of Mortality of New-York, April, 1819," *Medical Repository,* 20 (1820), 101; *Boston Medical and Surgical Journal,* 27 (1842 / 43), 125; [Mordecai M.] Noah, quoted in editorial in *Thomsonian Recorder,* 3 (1834 / 35), 127.

6. Samuel Thomson, *New Guide to Health; or, Botanic Family Physician,* 2d ed. (Boston: House, 1825).

7. Benjamin Waterhouse, letter to Samuel Thomson, in "Dr. Smith and Dr. Waterhouse," *Thomsonian Recorder,* 4 (1835 / 36), 125–126. The numbers of orthodox physicians with medical degrees who actually converted to botanical medicine prior to 1840 was apparently relatively few.

8. Samuel Thomson, *The Law of Libel: Report of the Trial of Dr. Samuel Thomson* (Boston: Lewis, 1839), p. 51.

9. "Every Man his Own Physician," *Southern Botanic Journal,* 1 (1837 / 38), 316–318.

10. B. W. [Benjamin Waterhouse], letter to the *Boston Courier,* Dec. 1834, reprinted in *Thomsonian Recorder,* 3 (1834 / 35), 155.

11. James S. Olcott, "Yellow Fever," *Southern Botanic Journal,* 3 (1839 / 41), 70. See also "Extraordinary Mortality," *Thomsonian Recorder,* 2 (1833 / 34), 48.

12. See "Report of Dr. J. H. Miller," *Thomsonian Recorder,* 4 (1835 / 36), 151–157.

13. For several such Thomsonian views, see "Keep Tally," *Botanico-Medical Recorder,* 10 (1841 / 42), 254; "Grave-Yard Statistics," ibid., p. 79; Wilson Thompson, letter to the editor, March 19, 1837, *Thomsonian Recorder,* 5 (1836 / 37), 235; "Preface," *Thomsonian Recorder,* 2 (1833 / 34), vii; "Scientific Doctors," *Southern Botanic Journal,* 3 (1839–1841), 151.

14. Gouverneur Emerson, "Medical Statistics," *American Journal of the Medical Sciences,* 1 (1827), 116.

15. By 1850, out of the thirty states and the District of Columbia,

only five (Connecticut, New Jersey, the District of Columbia, Louisiana, and Michigan) still had any licensing provisions. For a convenient summary, see "State Laws Respecting the Practice of Medicine," *Boston Medical and Surgical Journal,* 42 (1850), 109–113.

16. For typical accounts of the growth of quackery, see "Atmospheric Constitution and Diseases of New York," *Medical Repository,* 20 (1820), 210. [Daniel Drake], "From the Senior Editor," *Western Journal of Medicine and Surgery,* n.s. 4 (1845), 177. "Multiplication of Druggists and Apothecaries," *Boston Medical and Surgical Journal,* 39 (1848 / 49), 85–86; "Empiricism in Boston," ibid., 38 (1848), 305–307; "City Quacks," ibid., 41 (1849 / 50), 185; "Travelling Dentists," ibid., 29 (1843 / 44), 504. Joseph A. Eve, "Remarks on Empiricism," *Southern Medical and Surgical Journal,* 1 (1836 / 37), 70–75. "A Place without Quacks," *Boston Medical and Surgical Journal,* 35 (1846), 288.

17. "Dr. S. Thomson's Six Numbers," *Southern Botanic Journal,* 1 (1837 / 38), 138–139; "American Intelligence," *American Journal of the Medical Sciences,* n.s. 1 (1841), p. 272.

18. Samuel A. Cartwright, "Remarks on Statistical Medicine, contrasting the result of the empirical with the regular practice of Physic, in Natchez," *Western Journal of Medicine and Surgery,* 2 (1840), 1–21.

19. Thomas M. Logan, "An Essay on the beneficial results of Statistical Inquiries, in their application to Charleston, tracing the increased mortality to Quacks and Quackery," abstracted in *American Journal of the Medical Sciences,* n.s. 2 (1841), 245–248.

20. "American Intelligence," *American Journal of the Medical Sciences,* n.s. 1 (1841), 271–272; Charles Knowlton, "Quackery," *Boston Medical and Surgical Journal,* 34 (1846), 169–180.

21. "Grave-Yard Statistics," *Botanico-Medical Recorder,* 10 (1841 / 42), 79.

22. [Drake], "Travelling Editorials," *Western Journal of Medicine and Surgery,* 7 (1843), 471.

23. In 1802 Sinclair circulated to a few American correspondents a list of twenty questions on the factors affecting longevity. Although few replied, Sinclair did refer to several American publications in his *Code of Health and Longevity,* 4 vols. (Edinburgh: Constable; and London: Cadell, Davies, & Murray, 1807). In the United States, Sinclair's questionnaire was reprinted in the *Medical Repository,* 6 (1803), 351–352.

24. Harriet Martineau, *Society in America,* 4th ed., 2 vols. (New York: Saunders & Otley, 1837), II, 264; see also Bliss Perry, ed., *The Heart of Emerson's Journals* (Boston: Houghton Mifflin, 1926), p. 213.

25. John Bell, *On Regimen and Longevity* (Philadelphia: Haswell & Johnson, 1842), pp. 13–32, 377–420.

26. Horace Mann, "Report for 1842," in his *Annual Reports on Education* (Boston: Fuller, 1868), pp. 129–137, 229.

27. Sylvester Graham, *A Defence of the Graham System of Living* (New York: Applegate, 1835), pp. 16–27.

28. William A. Alcott, *Lectures on Life and Health* (Boston: Phillips & Sampson, 1853), pp. 20–46.

29. William A. Alcott, *The Physiology of Marriage* (Boston: Jewett, 1856), p. 119.

30. O. S. Fowler, *Love and Parentage* (New York: Fowlers and Wells, 1844).

31. Thomas Hersey, *The Midwife's Practical Directory*, 2d ed. (Baltimore: Thomas Hersey, 1836), p. 49.

32. "State Lunatic Hospital at Worcester, Mass.," *Boston Medical and Surgical Journal*, 12 (1835), 78; "Matrimonial Statistics," *Botanico-Medical Recorder*, 11 (1842 / 43), 74.

33. Orson S. Fowler, *Matrimony*, 61st ed. (New York: Fowlers & Wells, 1851). First issued in 1842, its subtitle was *Phrenology and Physiology applied to the Selection of Congenial Companions for Life*. See also Edward Jarvis, *Production of Vital Force* (Boston: Clapp, 1849), pp. 31–34.

34. "Deaths during the Winter," *Botanico-Medical Recorder*, 8 (1839 / 40), 285; ibid., 9 (1840 / 41), 285. A Boston editorial writer of 1846, commenting on the "vast number of persons wearing spectacles" in Boston, concluded that vanity and fashion were as much the cause as actual defective vision; see *Boston Medical and Surgical Journal*, 34 (1846), 365.

35. L. A. Quetelet, *Medical Examiner*, 1 (1838), 62–63.

36. Noah Webster, *A Brief History of Epidemic and Pestilential Diseases*, 2 vols. (Hartford: Hudson & Goodwin, 1799), II, 242–243; Charles Brockden Brown, footnote in his translation of C. F. Volney, *A View of the Soil and Climate of the United States of America* (Philadelphia: Conrad, 1804), 108–109.

37. John Bell, *On Baths and Mineral Waters* (Philadelphia: Porter, 1831), pp. 406–412.

38. Henry Wilson Lockette, *An Inaugural Dissertation on the Warm Bath* (Philadelphia: Carr & Smith, 1801); Charles Willson Peale, *An Epistle to a Friend, on the Means of Preserving Health, Promoting Happiness, and Prolonging the Life of Man to its Natural Period* (Philadelphia: Aitken, 1803). "Dr. Smith and Dr. Waterhouse," *Thomsonian Recorder*,

4 (1835 / 36), pp. 125–126. Harold Donaldson Eberlein reports that 401 Philadelphia homes had baths in 1823, and that beginning in 1829 plumbers were advertising both bathtubs and showertubs; see his "When Society First Took a Bath," *Pennsylvania Magazine of History and Biography*, 67 (1943), 46.

39. Martineau, *Society in America*, II, p. 260.

40. Isaac Parish, "Report on the Sanitary Condition of Philadelphia," *Transactions of the American Medical Association*, 2 (1849), 478; John Bell, *Report on the Importance and Economy of Sanitary Measures to Cities* (New York: Jones, 1859), p. 169.

41. Rush, "Inquiry into the State of Medicine in Philadelphia," pp. 234–237.

42. Webster, *History of Pestilential Diseases*, II, pp. 242–243.

43. Cooper, *The American Democrat*, pp. 154–157; Bell, *Regimen and Longevity*, p. 95.

44. Edward Jarvis, "Concord to Louisville 1837," MS journal, Jarvis Papers, Houghton Library, Harvard University.

45. Peale, *Epistle to a Friend*, passim.

46. William Beaumont, *Experiments and Observations on the Gastric Juice, and the Physiology of Digestion* (Plattsburgh, N.Y.: F. P. Allen, 1833).

47. E. J. [Edward Jarvis], review, *American Journal of the Medical Sciences*, n.s. 9 (1845), 389.

48. Bell, *Regimen and Longevity*, especially pp. viii, 389, 415.

49. Sylvester Graham, *Lectures on the Science of Human Life*, 2 vols. (Boston: Marsh, Capen, Lyon, & Webb, 1839), I, p. 459; II, pp. 36–41.

50. Graham, *Graham System of Living*, p. 18.

51. William A. Alcott, *Vegetable Diet*, 2d ed. (New York: Fowlers & Wells, 1849), p. 255.

52. E. J., "Vegetable Diet and Correct Regimen," *Boston Medical and Surgical Journal*, 15 (1836 / 37), 233–235. Alcott abstracted this in his *Vegetable Diet*, pp. 229–231. See also "The Graham System: Experiments in the Albany Orphan Asylum," *Water-Cure Journal*, 3–4 (1847), 89–93.

53. Alcott, *Vegetable Diet*, pp. 233–235, 263–269, 285.

54. William A. Alcott, *Tea and Coffee* (Boston: Light, 1839), pp. 163–173.

55. *Botanico-Medical Recorder*, 7 (1838 / 39), 364–365.

56. "Dr. Alcott's Work on Vegetable Diet," *Boston Medical and Surgical Journal*, 19 (1838 / 39), 253.

57. "Dietetics—Dr. Alcott's Work—No. II," *Boston Medical and Surgical Journal*, 19 (1838 / 39), 220; William A. Alcott, "Temperance and Excess," ibid., 21 (1839 / 40), 334–335; Samuel B. Woodward, "Medical Reminiscences—No. 1," ibid., 22 (1840), 10–11; John Bertram, "Medical Statistics," ibid., 41–43.

58. William A. Alcott, "Mortality in the United States," *Boston Medical and Surgical Journal*, 22 (1840), 46–47.

59. Charles A. Lee, "On Dietaries," *New York Journal of Medicine*, 1 (1843), 210–238, 341–369.

6. Quantitative Dimensions of Medical Sectarianism

1. As a group, these schools were so bad in the 1840s that few observers could have disagreed with the suggestion that the continued multiplication of their graduates could only "increase the number of those who will bring discredit upon the medical profession, and weaken its hold upon the confidence of the general public." I. F. Galloupe, "One Cause of Empiricism," *Boston Medical and Surgical Journal*, 41 (1849 / 50), 379–382.

2. By the Civil War there were fully a score of chartered botanical and eclectic schools alone; many others had been founded but had failed over the years. Most lists of medical schools that were compiled by regulars conspicuously omitted the irregular institutions.

3. Dan King, *Quackery Unmasked* (Boston: Clapp, 1858), pp. 332–334.

4. Ibid., p. 333. See also "City Quacks," *Boston Medical and Surgical Journal*, 41 (1849), 185; H. Graham, " 'Doctors' at the West," ibid., 50 (1854), 353–354; "Indian Doctors," ibid., 57 (1857), 82–83; "The treatment of Cancer—Cancer Quacks," ibid., 60 (1859), 225–226.

5. "Cincinnati Board of Health," *Boston Medical and Surgical Journal*, 40 (1849), 526.

6. M. M. Rodgers, "Medical Reform in Rochester, N.Y.—New Fee Bill," ibid., 50 (1854), 135–136; J. O. Harris, "The Medical Profession in the West," ibid., 49 (1853 / 54), 257–258.

7. Worthington Hooker, *Physician and Patient* (New York: Baker & Scribner, 1849), pp. 194–195.

8. James Harvey Young, "Quackery and the American Mind," *Cimarron Review*, 8 (1969), 36.

9. Bennet Dowler, "Natural History of Yellow Fever," *New Orleans Medical and Surgical Journal*, 15 (1858), 738; and "Researches into the Types of Disease and Types of Therapy," ibid., p. 594.

10. Elizabeth P. Peabody, *Memorial of Dr. William Wesselhoft* (Boston: Peabody, 1859), p. 10.

11. "Present Position of Medical Science," *American Homoeopathic Review*, 1 (1858 / 59), 2–4. See also M. M. Matthews, "Homoeopathy in Western New York," *American Journal of Homoeopathy*, 2 (1847 / 48), 134.

12. Constantin Hering, *A Concise View of the Rise and Progress of Homoeopathic Medicine* (Philadelphia: Hahnemannian Society, 1833), p. 21.

13. William H. Holcombe, *The Scientific Basis of Homoeopathy* (Cincinnati: Derby, 1852), passim.

14. J. H. Pulte, "Civilization and its Heroes: an Address," *Transactions of the American Institute of Homoeopathy*, 12 (1855), 42; F. Humphreys, "Homoeopathy in Western New-York," *American Journal of Homoeopathy*, 2 (1847 / 48), 123.

15. "Report of the Massachusetts Homoeopathic Medical Society," *Transactions of the American Institute of Homoeopathy*, 11 (1854), 48–49.

16. *Transactions of the American Institute of Homoeopathy*, 6 (1849), 4; ibid., 18 (1865), 23; "The Progression of Homoeopathia," *American Journal of Homoeopathy*, 3 (1849), 147; "An Argument against Homoeopathy!!!" ibid., 5 (1850 / 51), 107; "Homoeopathic Organization," *American Homoeopathic Review*, 6 (1865 / 66), 394. The new bureau was designed to maintain a register of physicians, compile lists of colleges, journals, hospitals, and societies, and collect other kinds of administrative information.

17. Constantin Hering, "On the Examination of the Sick," *American Journal of Homoeopathy*, 1 (1838 / 39), 118–122; C. Ticknor, "Letter," *Homoeopathic Examiner*, 1 (1840), 65, 69.

18. Peabody, *Memorial of Dr. William Wesselhoft*, pp. 21–27.

19. B. Fincke, "Homoeopathic Notation," *Transactions, American Institute of Homoeopathy*, 17 (1860), 117, 118, 122, 129.

20. Ibid. See also Dr. Lippe, letter to the editor, *American Journal of Homoeopathy*, 3 (1848 / 49), 43; "The High Potencies," ibid., pp. 104–105.

21. A. G. Hull, "The Editor's View of the State and Prospects of Homoeopathia," *Homoeopathic Examiner*, 1 (1840), 5–34; "Editorial Notices," ibid., p. 368; "Statistics of Homoeopathia in Europe," ibid., 2 (1841), 54–59.

22. P. M. Wheaton, letter to the editor, *American Journal of Homoeopathy*, 2 (1847 / 48), 194; William H. Holcombe, *Yellow Fever and its Homoeopathic Treatment* (New York: Radde, 1856), pp. 5, 40–42,

69–70; Daniel Holt and R. Shackford, "The Dysentery as it appeared in Lowell in 1847, Homoeopathically Treated," *Boston Medical and Surgical Journal*, 37 (1847 / 48), 479–482.

23. Clark Wright, "Homoeopathic Treatment of an Ophthalmia in the Protestant Half Orphan Asylum of this City," *Homoeopathic Examiner*, 3 (1842 / 43), 346–348; B. F. Bowers, "Reports in Relation to the Statistics and Medical Treatment at the New York Protestant Half Orphan Asylum for Seventeen Years," *American Journal of Homoeopathy*, 8 (1853 / 54), 17–29. Bowers's figures showed that the death rate had been over three times higher under allopathic treatment as under homeopathy.

24. E. Humphreys, "To the Inspectors of the State Prison . . . at Auburn," *Homoeopathic Examiner*, 3 (1842 / 43), 192. "Homoeopathic Prison Statistics," *Boston Medical and Surgical Journal*, 28 (1843), 502–503. Bushrod W. James, "Statistical Report of cases treated in the Northern Home for Friendless Children, Philadelphia, for seven and a half years, from the spring of 1857, to October 24, 1864, under Homoeopathic service," *Transactions, American Institute of Homoeopathy*, 18 (1865), 98–101. "Homoeopathy in Michigan," *American Homoeopathic Review*, 3 (1862 / 63), 378. B. F. Joslin, "Five Points House of Industry," ibid., 6 (1865 / 66), 142–146.

25. B. F. Joslin, "Lecture on Cholera," *Eclectic Medical Journal*, 8 (1849), 363–368.

26. B. F. Joslin, quoted in *American Journal of Homoeopathy*, 4 (1849 / 50), 11.

27. Ibid., p. 91; see also pp. 43–91, passim.

28. "Cholera," *Boston Medical and Surgical Journal*, 41 (1849 / 50), 20.

29. "Notices of Homoeopathic publications by our opponents of the Allopathic School," *American Journal of Homoeopathy*, 1 (1838 / 39), 143–144. See also "Miscellaneous," pp. 71–72; "Professor John P. Harrison of Cincinnati College and Homoeopathy," pp. 179–180.

30. Holcombe, *Yellow Fever*, p. 71; Holcombe, *The Scientific Basis of Homoeopathy*, p. ix; see also B. F. Bowers, review, *American Journal of Homoeopathy*, 2 (1847 / 48), 253.

31. John Forbes, *Homoeopathy, Allopathy and 'Young Physic'* (Philadelphia: Lindsay & Blakiston, 1846); James Jackson, *Another Letter to a Young Physician* (Boston: Ticknor & Fields, 1861); Jacob Bigelow, *Brief Exposition of Rational Medicine, to which is prefixed the Paradise of Doctors, a Fable* (Boston: Philips, Sampson, 1858); Oliver Wendell

Holmes, *Currents and Counter-Currents in Medical Science* (Boston: Ticknor & Fields, 1860).

32. Holcombe, *Scientific Basis of Homoeopathy,* p. 184.

33. J. H. Nutting, "On the Philosophy of Medical Delusions," *Boston Medical and Surgical Journal,* 49 (1853 / 54), 60–61. See also a review, *New York Journal of Medicine,* 6 (1846), 105–106; Worthington Hooker, *Homoeopathy: an Examination of its Doctrines and Evidences* (New York: Scribner, 1851), pp. 21–29.

34. Hooker, *Homoeopathy,* pp. 20–29. For varying treatments of the Lake Superior point and other details, see also: "Homoeopathy and its Thirtieth Dilution," *Saint Louis Medical and Surgical Journal,* 16 (1858), 474–475; Oliver Wendell Holmes, "Homoeopathy and its Kindred Delusions," in his *Medical Essays, 1842–1882* (Boston: Houghton Mifflin, 1889), pp. 42–45, 52–54; King, *Quackery Unmasked,* pp. 29–115, passim.

35. Holmes, "Homoeopathy and its Kindred Delusions," pp. 71, 76; James McCune Smith, "Lay Puffery of Homoeopathy," *Annalist,* 2 (1847 / 48), 348–351.

36. Editorial, "A Homoeopathic Hospital in St. Louis—its Statistics," *Saint Louis Medical and Surgical Journal,* 16 (1858), 279–280; see also L. Ch. B., "Homoeopathy—Its Standing in Europe—Its Statistics," ibid., pp. 145–149. Bennet Dowler, "Natural History of Yellow Fever," pp. 735–738. See also "Veracity of Homoeopathists," *Buffalo Medical Journal,* 4 (1848 / 49), 194–196.

37. Hooker, *Homoeopathy,* pp. 107–111.

38. Holcombe, *Scientific Basis of Homoeopathy,* p. 222; C. Ticknor, letter, *Homoeopathic Examiner,* 1 (1840), 62–72; Holmes, "Homoeopathy and its Kindred Delusions," pp. 60–61, 79–80: Elisha Bartlett, *An Inquiry into the Degree of Certainty in Medicine; and into the Nature and Extent of its Power over Disease* (Philadelphia: Lea & Blanchard, 1848), p. 80. See also Dowler, "Natural History of Yellow Fever," p. 26.

39. [Constantine] Rafinesque, *Medical Flora; or, Manual of the Medical Botany of the United States of North America,* 2 vols. (Philadelphia: Atkinson & Alexander, 1828), I, i–vii.

40. [Constantine] Rafinesque, *The Pulmist* (Philadelphia: C. Alexander, 1829).

41. "Proceedings of Ohio State Eclectic Medical Association, 1852," *Eclectic Medical Journal,* 12 (1853), 53.

42. T. V. Morrow, "Chancellor Curtis' Calculation," *Western Medical Reformer,* 5 (1845/46), 139–143.

43. J. M. Scudder, "A Brief History of Eclectic Medicine," *Eclectic Medical Journal*, 39 (1879), 297–308.

44. "Statistics of Calomel," *Western Medical Reformer*, 5 (1845/46), 52–53; "Statistics of Medical Schools," *Eclectic Medical Journal*, 13 (1854), 363–364.

45. "Proceedings of Ohio State Eclectic Medical Association, 1852," pp. 53–56; *Eclectic Medical Journal*, 8 (1849), 245–247.

46. "Illustrations of Eclectic Practice," *Eclectic Medical Journal*, 8 (1849), 542–545; "Cholera in Cincinnati—Triumph of Liberal Principles," ibid., pp. 284–288.

47. Victor J. Fourgeaud, "Eclecticism in Medicine," *Saint Louis Medical and Surgical Journal*, 3 (1845 / 46), 98–109; Hiram Cox, letter to the editor, *Western Medical Reformer*, 5 (1845 / 46), 101–103; Joseph R. Buchanan, "American Eclecticism: an Introductory Lecture (Nov. 4, 1850)," *Eclectic Medical Journal*, 9 (1850), 489–491.

48. John E. M'Geer, "Inoculation in Rubeola," *Eclectic Medical Journal*, 10 (1851), 322–324.

49. The editor of the *Boston Medical and Surgical Journal* felt that Priessnitz's success as a medical practitioner had been abundantly proved by the three main yardsticks that society was accustomed to use during the 1840s: 1) numbers of patients; 2) numbers of notable patients; and 3) accumulated wealth. Editorial, *Boston Medical and Surgical Journal*, 27 (1842), 283.

50. Henry C. Wright, letters to William Lloyd Garrison, "Hydropathy—or the Water Cure," *Western Medical Reformer*, 4 (1844), 17–19.

51. Ibid., pp. 49–52; Mrs. M. L. Shew, *Water-Cure for Ladies* (New York: Wiley & Putnam, 1844), p. vii.

52. Joel Shew, *Hand-Book of Hydropathy* (New York: Wiley & Putnam, 1844), pp. v–vi; Mrs. M. L. Shew, *Water-Cure for Ladies*, passim; G. W. Bourne, "The Science of Medicine," *Water-Cure Journal*, 12 (1851), 12.

53. R. T. Trall, *The Hydropathic Encyclopedia*, 2 vols. (New York: Fowlers & Wells, 1854), I, 42; Joel Shew, *The Cholera* (New York: Fowlers & Wells, 1855), pp. 29, 70–71, 84–85.

54. "Dr. Shieferdecker's Preface to his Cases," in *The Water-Cure in America* [ed. H. F. Phinney] (New York: Fowlers & Wells, 1856), pp. 9–10.

55. Trall, *Hydropathic Encyclopedia*, I, 382–385, and II, 439, 446; and "Topics of the Month," *Water-Cure Journal*, 32 (1861), 129. See also "Increasing Demand for Practitioners of Water-Cure," *Water-Cure Journal*, 7 (1849), 178.

56. "Statistics from the American Water-Cure Establishments," *Water-Cure Journal,* 12 (1851), 89–90.

57. "Statistics—a Letter," *Water-Cure Journal,* 19 (1855), 35; "Round Hill Water Cure and Motorpathic Institute, at Northampton, Mass.," ibid., p. 91.

58. R. T. Trall, "March Matters," *Water-Cure Journal,* 15 (1853), 61–62; and "July Cogitations," ibid., 12 (1851), 13–14. See also Joel Shew, *Children: their Hydropathic Management in Health and Disease* (New York: Fowlers & Wells, 1852), pp. v–vi; Mrs. M. L. Shew, *Water-Cure for Ladies,* pp. 14–15.

59. J. H. Nutting, "On the Philosophy of Medical Delusions," pp. 55–56. See also "Report of 563 Cases Treated at Brattleboro," *Boston Medical and Surgical Journal,* 39 (1848 / 49), 84–85.

60. "Increase of Hydropathic Hospitals," *Boston Medical and Surgical Journal,* 37 (1847), 325. See also "The Hydropathists—their Water-Cure Almanac," ibid., 48 (1853), 506; "Hydropathic Medical College," ibid., 52 (1855), 61.

61. [Austin Flint], "Sanitary Institutions," *Buffalo Medical Journal,* 6 (1850 / 51), 753–754. John Bell, "Mineral and Thermal Springs of the United States," *Medical Examiner,* 11 (1855), 333–342, 394–402.

7. Statistics of Mind and Madness

1. Amariah Brigham, *Remarks on the Influence of Mental Cultivation and Mental Excitement upon Health,* 2d ed. (Boston: Marsh, Capen, & Lyon, 1833), pp. iv, 82–83.

2. These antecedents were well understood by contemporary observers. See "Statistics of Insanity," *American Journal of Insanity,* 18 (1861), 1–14.

3. [Daniel Drake], "Travelling Editorials," *Western Journal of Medicine and Surgery,* 8 (1843), 76–77; "From the Senior Editor," *Western Journal of Medicine and Surgery,* n.s. 4 (1845), 178; Josiah C. Nott, "A Lecture on Animal Magnetism," *Southern Journal of Medicine and Pharmacy,* 1 (1846), pp. 261–289.

4. John Kearsley Mitchell, "An Essay on Animal Magnetism, or Vital Induction," in his *Five Essays,* ed. S. Weir Mitchell (Philadelphia: Lippincott, 1859), pp. 143–274.

5. Charles Dickens, *American Notes,* ed. Marcus Stone et. al. (London: Oxford University Press, 1957), pp. 180–182. See also [J. V. C. Smith?], "Statistics of Phrenology in the United States," in *The*

American Medical Almanac, for 1841, ed. J. V. C. Smith (Boston: Otis, Broaders, 1841), pp. 117–120.

6. Joseph R. Buchanan, *Outlines of Lectures on the Neurological System of Anthropology* (Cincinnati: Buchanan's Journal of Man, 1854), pp. 81, 276–277. Levi Reuben, "Phrenology; its History and Doctrines—No. 2," *American Phrenological Journal,* 34 (1861), 2–4; Charles Caldwell, *Elements of Phrenology* (Lexington, Ky.: Meriwether, 1827), pp. 95–98; O. S. Fowler and L. N. Fowler, *Phrenology Proved, Illustrated, and Applied* (New York: Colyer, 1836), pp. 43–51.

7. Fowler and Fowler, *Phrenology Proved,* pp. iv, 43–44; O. S. Fowler, *Fowler on Memory: or Phrenology Applied to the Cultivation of Memory* (New York: Fowler, 1842), pp. 8–13, 62–63.

8. [Smith], "Statistics of Phrenology," pp. 117–120.

9. Stanton found that the only American collections that were used for serious craniological research were those of Morton, Buchanan, and John C. Warren. William Stanton, *The Leopard's Spots* (Chicago: University of Chicago Press, 1960), p. 29.

10. Samuel G. Morton, *An Illustrated System of Human Anatomy, Special, General and Microscopic* (Philadelphia: Grigg & Elliot, 1849), pp. 69–72.

11. Samuel G. Morton, *Crania Americana, or a Comparative View of the Skulls of Various Aboriginal Nations of North and South America* (Philadelphia: Dodson; and London: Simpkin & Marshall, 1839). See also J. Aitken Meigs, *Catalogue of Human Crania, in the Collection of the Academy of Natural Sciences of Philadelphia based upon the Third Edition of Dr. Morton's "Catalogue of Skulls"* (Philadelphia: Merrihew and Thompson, 1857).

12. Editorial, "Study of Ancient Crania," *Boston Medical and Surgical Journal,* 31 (1844 / 45), 422–423.

13. "Mr. Webster's Brain," *Western Journal of Medicine and Surgery,* 3d ser., 11 (1853), 92; John Jeffries, "An Account of the last Illness of the Honorable Daniel Webster," *American Journal of the Medical Sciences,* n.s. 25 (1853), 116–119.

14. Samuel Gridley Howe, *An Address Delivered at the Anniversary Celebration of the Boston Phrenological Society, December 28th, 1835* (Boston: Marsh, Capen, & Lyon, 1836), pp. 10, 15.

15. Elisha Bartlett, "An Address, delivered at the Anniversary Celebration of the birth of Spurzheim and of the Organization of the Boston Phrenological Society, January 1, 1838," *American Phrenological Journal,* 1 (1839), 148–155.

16. L. A. J. Quetelet, *Sur l'homme et le développement de ses facultés,*

ou, Essai de physique sociale (Brussels: Hauman, 1836). An English translation was published in 1842.

17. L. A. J. Quetelet, *Anthropometrie* (Brussels: Muquardt; and Paris: Baillière, 1871), pp. 315–322. See also the L. A. J. Quetelet Letters in the library of the Académie Royale de Belgique (microfilm at the American Philosophical Society): Thomas G. Clemson to George Catlin, undated; George Catlin to Quetelet, undated; Robley Dunglison to Quetelet, June 24, 1849, and Aug. 29, 1849.

18. Timothy Flint, *Recollections of the Last Ten Years* (Boston: Cummings, Hilliard, 1826), pp. 40–41, 240–241, 251, 311.

19. "Fast Living," *Boston Medical and Surgical Journal*, 55 (1856 / 57), 454; Henry Ward Beecher, "Physical Culture," *American Phrenological Journal*, 32 (1860), 32.

20. J. Edwards Lee, "Statistics of the Suicides which have occurred in the State of New York, from Dec. 1, 1844, to Dec. 1, 1846," *American Journal of Insanity*, 3 (1846 / 47), 349–352; E. K. Hunt, "Statistics of Suicides in the United States," ibid., 1 (1844 / 45), 225–234.

21. Beecher, "Physical Culture," p. 32. See also "Money-Making Mania," *American Journal of Insanity*, 5 (1848 / 49), 327–328; P. E. [Pliny Earle], review, *American Journal of the Medical Sciences*, n.s. 37 (1859), 502; "Public Amusements," *Boston Medical and Surgical Journal*, 49 (1853 / 54), 247.

22. Amariah Brigham, *The Influence of Mental Cultivation*, pp. iv, 82–83.

23. Amariah Brigham, *Observations on the Influence of Religion upon the Health and Physical Welfare of Mankind* (Boston: Marsh, Capen, & Lyon, 1835), pp. 139–150.

24. [Amariah Brigham?], "Millerism," *American Journal of Insanity*, 1 (1844 / 45), 252.

25. "Moral Thermometer," *Journal of Health*, 4 (1833), p. 5.

26. The Pennsylvania Hospital had admitted the insane from its beginning in 1752, while in 1770 the Williamsburg, Virginia, asylum was launched as the first institution in this country specifically devoted to the mentally ill.

27. T. Romeyn Beck, "Account of some of the Lunatic Asylums in the United States," *New-York Medical and Physical Journal*, 7 (1828), 186–206.

28. "Lunatics in the United States," *Boston Medical and Surgical Journal*, 12 (1835), 34.

29. The controversy over this census is discussed later in this chapter.

30. Pliny Earle, *A Visit to Thirteen Asylums for the Insane in Europe* (Philadelphia: Dobson, 1841). See also "Lunatic Asylums of the United States," *American Journal of Insanity*, 2 (1845 / 46), 46–48.

31. Dorothea Dix, *Memorial Soliciting a State Hospital for the Insane, Submitted to the Legislature of New Jersey, January 23, 1845*, 2d ed. (Trenton, 1845), p. 4; *Memorial to the Honorable the Legislature of the State of New York* (Albany, 1844), p. 57; and *Memorial of Miss D. L. Dix to the Honorable the General Assembly in Behalf of the Insane of Maryland*, Senate Document C (Annapolis, 1852), pp. 3–8.

32. The McLean Asylum, for one, as early as 1825 was keeping four different patient record books: 1) therapies used, 2) drug prescriptions, 3) individual case records, and 4) pulse records. See the diary of George W. Folsom, in Nina Fletcher Little, *Early Years of the McLean Hospital* (Boston: Countway Library of Medicine, 1972).

33. *Eighth Annual Report of the Trustees of the State Lunatic Hospital at Worcester, December 1840* (Boston: Dutton & Wentworth, 1841), p. 60.

34. *Sixth Annual Report of the Trustees of the State Lunatic Hospital at Worcester, December 1838* (Boston: Dutton & Wentworth, 1839), pp. 7–10.

35. "State Lunatic Hospital Reports," *Boston Medical and Surgical Journal*, 17 (1837), 64–65.

36. *Tenth Report of the Directors of the Connecticut Retreat for the Insane, May, 1834* (Hartford: Hudson & Skinner, 1834), p. 9, *Worcester Asylum, Eighth Annual Report, 1840*, pp. 58–68; Pliny Earle, "On the Curability of Insanity," *American Journal of the Medical Sciences*, n.s. 5 (1843), 350.

37. Earle, "On the Curability of Insanity," pp. 350–351.

38. Thomas Kirkbride, "Report for 1845," *Reports of the Pennsylvania Hospital for the Insane* (Philadelphia: Pennsylvania Hospital for the Insane, 1846), pp. 10–11.

39. Pliny Earle, "Researches in reference to the Causes, Duration, Termination, and Moral Treatment of Insanity," *American Journal of the Medical Sciences*, 22 (1838), 355. See also his "Curability," pp. 347, 352–358; *History, Description, Statistics of the Bloomingdale Asylum for the Insane* (New York: Egbert, Hovey, & King, 1848), p. 130; *Visit to Thirteen Asylums*, passim; and his "The Curability of Insanity," *The Annalist*, 2 (1847 / 48), 127–130.

40. Luther Bell, "Twenty-fourth annual report of the physician and superintendent of the McLean Asylum for the Insane, to the Trustees of the Massachusetts General Hospital," in *Annual Report*

of the Board of Trustees of the Massachusetts General Hospital for the Year 1841 (Boston: Loring, 1842), p. 12.

41. Isaac Ray, "The Statistics of Insane Hospitals," *American Journal of Insanity,* 6 (1849), 23–52.

42. "Statistics of Insanity," by the editor [Amariah Brigham], *American Journal of Insanity,* 6 (1849), 141–145. See also: Richard J. Dunglison, "Statistics of Insanity in the United States," *North American Medical and Chirurgical Review,* 4 (1860), 656, 693. "Statistics of Insanity," *American Journal of Insanity,* 18 (1861), 12–14.

43. Edward Jarvis, manuscript journal, "Concord to Louisville 1837," p. 92, in the Jarvis Papers, Harvard University.

44. E. J. [Jarvis], review, *Western Journal of Medicine and Surgery,* 4 (1841), 449, 476–477.

45. Edward Jarvis, "What shall we do with our Insane?," *Western Journal of Medicine and Surgery,* 5 (1842), 82–83, 116–117.

46. E. J., review, *Western Journal of Medicine and Surgery,* 4 (1841), 472–477. "Insanity in Kentucky," *Boston Medical and Surgical Journal,* 24 (1841), 169–171; E. J., "Prospects of the Blind and Insane in Kentucky," ibid., 26 (1842), 60–61.

47. Edward Jarvis, "Statistics of Insanity in the United States," *Boston Medical and Surgical Journal,* 27 (1842), 116–121, 281–282. See also "Reflections on the Census of 1840," *Southern Literary Messenger,* 9 (1843), 340–353; Jarvis, "Insanity among the Coloured Population of the Free States," *American Journal of the Medical Sciences,* n.s. 7 (1844), 71–83; Jarvis, "Insanity among the Coloured Population of the Free States," *American Journal of Insanity,* 8 (1851 / 52), 281 ff.

48. Edward Jarvis, "On the Comparative Liability of Males and Females to Insanity, and their Comparative Curability and Mortality when Insane," *American Journal of Insanity,* 7 (1850 / 51), 142–171.

49. Only Belgium and Norway, Jarvis noted, had up to then conducted even one careful and reliable enumeration of the insane.

50. Edward Jarvis, "On the Supposed Increase of Insanity," *American Journal of Insanity,* 8 (1851 / 52), 333–364.

51. Gerald N. Grob, ed., *Insanity and Idiocy in Massachusetts: Report of the Commission on Lunacy, 1855, by Edward Jarvis* (Cambridge, Mass.: Harvard University Press, 1971), passim. Review, *Boston Medical and Surgical Journal,* 52 (1855), 321–323; review, *American Journal of the Medical Sciences,* n.s. 31 (1856), 429–436.

52. Edward Jarvis, "Distribution of Lunatic Hospital Reports," *American Journal of Insanity,* 14 (1857 / 58), 248–253.

53. Among the members of this group, Grob identified Jarvis, John H. Griscom, Edwin M. Snow, Edward Barton, Elisha Harris, Wilson Jewell, and Henry G. Clark. Gerald N. Grob, *Edward Jarvis and the Medical World of Nineteenth-Century America* (Knoxville: University of Tennessee Press, 1978), pp. 7, 63. To his list should be added such names as John Bell, E. N. Fenner, Elisha Bartlett, and the Paris-educated opponents of heroic therapeutics, among others.

54. Edward Jarvis, *Production of Vital Force* (Boston: Clapp, 1849), p. 7.

55. Ibid., p. 2.

8. Vital Statistics and Public Health

1. For a review of vital statistics provisions immediately after Independence, see Chapter I.

2. In 1855, Edward Jarvis reported that the following American cities had published their bills, "some for many years": Lowell, Boston, Roxbury, Providence, New York, Brooklyn, Philadelphia, Baltimore, Washington, Charleston, New Orleans, Memphis. E. J. [Jarvis], review, *American Journal of the Medical Sciences*, n.s. 29 (1855), 408; see also Jarvis to George Graham, June 10, 1853, Jarvis Letterbook 3, Jarvis Papers, Harvard University.

3. Charles Caldwell, *Medical and Physical Memoirs* (Philadelphia: Bradford, 1801), pp. 44–46; John Bell, *Report on the Importance and Economy of Sanitary Measures to Cities* (New York: Jones, 1859), pp. 180–197; Edward Jarvis, review in *American Journal of the Medical Sciences*, n.s. 9 (1845), 131–154; Timothy Dwight, *Travels in New England and New York*, 4 vols. in 2, ed. Barbara Miller Solomon (Cambridge, Mass.: Harvard University Press, 1969), II, 60; Benjamin Henry Latrobe, *The Journal of Latrobe* (New York: Appleton, 1905), pp. 194–201.

4. Felix Pascalis, *An Exposition of the Danger of Interment in Cities* (New York: Gilley, 1823); Bell, *Importance of Sanitary Measures*, p. 196; "Awful Calculation," *Thomsonian Recorder*, 2 (1833 / 34), 222.

5. Andrew Jackson Downing, *Rural Essays* (New York: Putnam, 1853), pp. 154–155; Jacob Bigelow, "On the Burial of the Dead," in his *Nature in Disease* (Boston: Ticknor & Fields, 1854); Cornelia Walter, *Mount Auburn Illustrated* (New York: Martin, 1847).

6. Latrobe, *Journal*, pp. 241–242; see also Caldwell, *Memoirs*, p. 12.

7. Latrobe, *Journal,* pp. 275–300; Bennet Dowler, *Tableau of the Yellow Fever of 1853* (New Orleans: Picayune, 1854), pp. 29–34, 45–46.

8. *Boston Medical and Surgical Journal,* 17 (1837), 209–210. See also Walter Channing, "Bills of Mortality," *New England Journal of Medicine and Surgery,* 15 (1826), 225–234; "Bills of Mortality," *Boston Medical and Surgical Journal,* 1 (1828), 109–111, 205–207; ibid., 14 (1836), 32–33; ibid., 16 (1837), 35, 208–209.

9. The City Inspector of New York, who was the preparer during this time (1816–1829), was Dr. George Cuming. While nonphysicians occupied this office for several years both before and after Cuming, a series of physicians again filled it beginning in 1835 and to the midforties. Philadelphia's mortality data were also compiled under medical supervision.

10. Nathaniel Niles and John D. Russ, "Medical Statistics; or a Comparative View of the Mortality in New-York, Philadelphia, Baltimore, and Boston, for a Series of Years," *New York Medical and Physical Journal,* 6 (1827), 304–312. The article was also published as a pamphlet (New York: Bliss, 1827).

11. Unsigned review, *Boston Medical and Surgical Journal,* 18 (1838), 97–98. See also D. Humphreys Storer, "Medical Statistics and Bills of Mortality, for Boston, during Nineteen Years ending January 1st, 1832," *Medical Magazine,* 1 (1833), 509–529, 576–598; Charles A. Lee, "Medical Statistics," *American Journal of the Medical Sciences,* 19 (1836 / 37), 25–52.

12. Gouverneur Emerson, "Medical Statistics," *American Journal of the Medical Sciences,* 1 (1827), 116–117; G. E. [Emerson], "New Form of Certificate Required by the Philadelphia Board of Health," *American Journal of the Medical Sciences,* 20 (1837), 535–536.

13. However, physician members of medical societies did play active roles as advisors to boards of health on sanitary measures, on the distribution of information to the public, on the organization of emergency hospitals, and related matters.

14. John Bell and D. Francis Condie, *All the Material Facts in the History of Epidemic Cholera: Being a Report of the College of Physicians of Philadelphia to the Board of Health* (Philadelphia: Desilver, 1832), especially pp. 84–142; "Report of the Committee on Practical Medicine and Epidemics," *Transactions of the American Medical Association,* 3 (1850), 113–114.

15. Among the varied works appearing in 1832 were H. S. Tan-

ner, *A Geographical and Statistical Account of the Epidemic Cholera* (Philadelphia: Tanner, 1832); and Amariah Brigham, *A Treatise on Epidemic Cholera* (Hartford: Huntington, 1832).

16. *Boston Medical and Surgical Journal,* 5 (1832), 416; "Works on the Cholera," *Boston Medical and Surgical Journal,* 7 (1833), 186.

17. For examples of this statistical variety see *American Journal of the Medical Sciences,* 11 (1832 / 33), 268–273.

18. Samuel Jackson, "Personal Observations and Experience of Epidemic or Malignant Cholera in the City of Philadelphia," *American Journal of the Medical Sciences,* 11 (1832 / 33), 289–346. See also Calvin Jewett, "Statistical Account of Cholera in America," *Boston Medical and Surgical Journal,* 7 (1832), 64–65; "March of the Cholera," ibid., p. 253.

19. Daniel Drake, *An Account of the Epidemic Cholera as it Appeared in Cincinnati* (Cincinnati: Deming, 1832), pp. 17, 20.

20. "Report of the Committee on Practical Medicine and Epidemics," *Transactions of the American Medical Association,* 3 (1850), 107–114. See also review, *New York Journal of Medicine,* n.s. 5 (1850), 241–243; William P. Buel, "Remarks on the Asiatic-Cholera, in the City of New-York in 1848–9," *New York Journal of Medicine,* n.s. 4 (1850), 25–27.

21. D. F. C. [Condie], review, *American Journal of the Medical Sciences,* n.s. 18 (1849), 437–443; see also "Statistics of Cholera," *Buffalo Medical Journal,* 12 (1856 / 57), 317–319.

22. "Address of the Eclectic Medical Society of Cincinnati to the People of the United States," *Eclectic Medical Journal,* 8 (1849), 337–343; "Statistics of Cholera Practice," ibid., pp. 314–316; "Value of Electicism," ibid., pp. 571–572.

23. Buel, "Remarks on the Cholera," pp. 15–16.

24. "The Cholera—its Course and Ravages," *Boston Medical and Surgical Journal,* 41 (1849 / 50), 123; "Cholera in St. Louis, Mo.," ibid., 45 (1854 / 55), 84; Buel, "Remarks on the Cholera," p. 17; William M. McPheeters, "History of Epidemic Cholera in St. Louis in 1849," *Saint Louis Medical and Surgical Journal,* 8 (1850), 108.

25. James W. Stone, "The Cholera Epidemic in Boston," *Boston Medical and Surgical Journal,* 41 (1849 / 50), 298–299; Lemuel Shattuck, *Report to the Committee of the City Council appointed to obtain the Census of Boston for the Year 1845* (Boston: Eastburn, 1846), pp. 31, 129. The only city then known to have a district of greater population density was Liverpool.

26. "Cincinnati Board of Health," *Boston Medical and Surgical Journal,* 40 (1849), 525–526; "Epidemic Cholera in Philadelphia," *Medical*

Examiner, n.s. 5 (1849), 615–616. For examples of the irregulars' own statistics, some of which they viewed as proving the "merits of the Eclectic Reform" over regular practice, see B [Joseph R. Buchanan], "Cholera in Cincinnati—Triumph of Liberal Principles," *Eclectic Medical Journal*, 8 (1849), 284–288. For homeopathic statistics, see "Cholera," *Boston Medical and Surgical Journal*, 41 (1849 / 50), 18–21.

27. Buel, "Remarks on the Cholera," p. 16; McPheeters, "Cholera in St. Louis," pp. 102, 105. "Statistics and Reports on Cholera," *Boston Medical and Surgical Journal*, 41 (1849 / 50), 383–384; J. C. Simonds, Statistical Researches on the Epidemic Cholera which prevailed in New-Orleans, from December 12th, 1848, to June, 1849," *Charleston Medical Journal*, 4 (1849), 566–567.

28. Unsigned editorial, presumably by Silliman, "New York Statistical Society," *American Journal of Science*, 32 (1837), 203; note [Silliman?], ibid., 34 (1838), 213–214.

29. James Fenimore Cooper, *Notions of the Americans*, 2 vols. (New York: Stringer & Townsend, 1850), I, 286.

30. Archibald Russell, *Principles of Statistical Inquiry* (New York: Appleton, 1839).

31. See, for instance, Gouverneur Emerson, "Review," *American Journal of the Medical Sciences*, 20 (1837), 462; and editorial, "The Science of Statistics," *DeBow's Review*, 3 (1847), 270. The emergence of medical almanacs was epitomized by the highly statistical *American Medical Almanac*, issued in Boston beginning in 1839 by J. V. C. Smith. But regular medical journals, too, devoted extraordinary amounts of space to original statistical articles, abstracts of foreign statistical works, summaries, and reviews of statistical reports. See, for instance, *American Journal of the Medical Sciences*, n.s. 1 (1841).

32. Charles Sanderson, "On the establishment of Statistical Societies in the United States," *American Journal of Science*, 31 (1836 / 37), 186–188. "New York Statistical Society," *American Journal of Science*, 32 (1837), 202–203. John K. Wright, *Geography in the Making: the American Geographic Society, 1851-1951* (New York: American Geographic Society, 1952). *Constitution and By-Laws of the Statistical Society of Pennsylvania* (Philadelphia: Crissy & Markley, 1847). Samuel G. Morton was vice president of the Pennsylvania society in 1847, and Gouverneur Emerson a council member, but few other Philadelphia physicians appeared on the list of members (*Constitution and By-Laws*, pp. 21–24). In the American Geographical and Statistical Society, physicians apparently did not play an important role. That society's early leaders included the educator and publicist Archibald Russell,

the clergyman Joshua Leavitt, and the editor-economist Henry Varnum Poor.

33. Lemuel Shattuck, "Circular," dated April 4, 1840, reprinted in "Proceedings of the Centenary Celebration, 1839–1939," *Journal of the American Statistical Association*, 35, pt. II (1940), p. 303. The subsequent discussion of the association is based largely upon archival materials reproduced in this work. Of the secondary accounts the most useful is Walter F. Willcox, "Lemuel Shattuck, Statist, founder of the American Statistical Association," ibid., p. 224–235.

34. Ideologically, the membership thus included social conservatives as well as social activists, ardent advocates of laissez faire as well as those who took statistics literally as a discipline committed to extend the role of the state.

35. In addition to the fourteen local physicians who were original Fellows, two American physicians (Samuel G. Morton and John W. Francis) were among the thirty-six honorary members, and three physicians (Robley Dunglison, Gouverneur Emerson, and Jarvis) were among fifteen American corresponding members.

36. Shattuck, *A History of the Town of Concord, Middlesex County, Massachusetts* (Boston: Russell, Odiorne, 1835).

37. Shattuck to Quetelet, Dec. 29, 1839, L. A. J. Quetelet Letters in the Académie Royale de Belgique, Brussels (microfilm copy at the American Philosophical Society). The medical almanac typically included statistics of medical colleges, hospitals, dispensaries, journals, societies, and prizes, along with posological tables, common prescriptions, and historical dates. Smith's almanacs of 1840 and 1841 also included short articles, filled with statistics, on specific medical topics.

38. Lemuel Shattuck, "On the Vital Statistics of Boston," *American Journal of the Medical Sciences*, n.s. 1 (1841), 369–401.

39. Lemuel Shattuck to John A. Bolles, Dec. 13, 1843, in Bolles, *Second Annual Report to the Legislature, under the Act of March, 1842, Relating to the Registry and Returns of Births, Marriages and Deaths in Massachusetts, for the year ending May 1st 1843* (Boston: Dutton & Wentworth, 1843), pp. 64–86; Shattuck, "Letter to the Secretary," Dec. 12, 1845, appendix to John G. Palfrey, *Fourth Annual Report to the Legislature, relating to the Registry and Returns of Births, Marriages, and Deaths in Massachusetts, for the year ending April 30th, 1845* (Boston: Dutton & Wentworth, 1845), pp. 67–106. Shattuck outlined additional improvements for the system in communications to the state legislature in 1848 and 1849. For a comment on the influence of

the Massachusetts initiative, see review, *New York Journal of Medicine,*
3 (1844), 66.

40. Shattuck, "Letter to the Secretary," pp. 91–99.

41. Edward Jarvis, review, *American Journal of the Medical Societies,*
n.s. 9 (1845), 387–388.

42. Shattuck, "Letter to the Secretary," pp. 91–99.

43. Franklin Tuthill, "Registration of Births, Deaths, and Marriages," *Transactions of the Medical Society of the State of New York* (1853),
12–13.

44. See Gerald N. Grob, *Edward Jarvis and the Medical World of
Nineteenth-Century America* (Knoxville: University of Tennessee Press,
1978).

45. *Statistical Report on the Sickness and Mortality in the Army of the
United States . . . from January, 1819, to January, 1839* (Washington:
Gideon, 1840). Forry's name does not appear on the title page of
this work. His important contribution to the statistical analysis of
meteorology and medical topography are examined in another volume.

46. J. S., "The Late Samuel Forry," *New York Journal of Medicine,*
4 (1845), 7–10; Charles A. Lee, "Report of the Standing Committee
of the Society on Hygiene and Medical Statistics," *Transactions of the
Medical Society of the State of New York,* 1850, p. 151; and Charles A.
Lee to Lemuel Shattuck, Jan. 17, 1846, Shattuck Papers, Massachusetts Historical Society.

47. *Boston Medical and Surgical Journal,* 24 (1841), 179; Massachusetts Secretary of the Commonwealth, *First Annual Report to the Legislature, under the Act of March, 1842, Relating to the Registry and Returns
of Births, Marriages and Deaths in Massachusetts* (Boston, 1843), pp.
1–27.

48. Bennet Dowler, review, *New Orleans Medical and Surgical Journal,* 8 (1851 / 52), 510.

49. See *Proceedings of the National Medical Conventions, held in New
York, May, 1846, and in Philadelphia, May, 1847* (Philadelphia: Collins,
1847), pp. 20–21, 34, 43, 125–131; and letters from Griscom to
Shattuck, Shattuck Papers, Henry J. Huntington Library.

50. [Lemuel Shattuck et al.], *Report of a General Plan for the Promotion of Public and Personal Health . . . of Massachusetts . . .* (Boston:
Dutton & Wentworth, 1850), pp. 149–150.

51. For the committee's formation and report, see *Proceedings,
National Medical Conventions,* pp. 21, 35, 37, 133–175. The committee's list of 1,147 names for diseases then in actual use in the United

States, which took up some thirty pages, was compiled from the bills of Baltimore, Boston, Charleston, New York, Philadelphia, and rural Massachusetts. See also the John H. Griscom letters to Shattuck, Shattuck Papers, Henry J. Huntington Library. In 1860 a number of different nosologies were used in compiling state and city registration reports. However, variants of the Farr classification had been adapted in mortality reports of the United States Army, the Commonwealth of Massachusetts, and the 1850 and 1860 censuses, among others. See [U.S. Bureau of the Census], *Eighth Census of the United States, 1860* (Washington: Government Printing Office, 1866), pp. xxvi–xxviii.

52. These states were Massachusetts, Vermont, Connecticut, Rhode Island, New Jersey, South Carolina, and Kentucky. [U.S. Bureau of the Census], *Statistics of the United States (including Mortality, Property, &c) in 1860* (Washington: Government Printing Office, 1866), p. xxv; see also E. J. [Edward Jarvis], "Review," *American Journal of the Medical Sciences*, n.s. 29 (1855), 407–410; Jarvis, "Report on Registration of Births, Marriages, and Deaths," *Transactions of the American Medical Association*, 11 (1858), 527–535; Josiah Curtis, "On the System of Registration in the United States of America," *Journal of the Statistical Society of London*, 17 (1854), 43–44.

53. E. J. [Edward Jarvis], "Review," *American Journal of the Medical Sciences*, n.s. 29 (1855), 410.

54. State of California, *Annual Report of the State Registrar for the Year 1859* (Sacramento, 1860), p. 3; *Letters of Richard D. Arnold, M.D., 1808-1876*, ed. Richard H. Shryock (New York: AMS Press, 1970), pp. 31–33, 36–37, 39.

55. William L. Sutton in 1859 and 1860 saw that the system he had organized in Kentucky was "doomed . . . in part owing to a want of satisfaction with the [meagerness] of the returns, & in part to troublous times." Sutton to Edward Jarvis, June 19, [1860], Jarvis Papers, Harvard University. For an account of the slow progress of registration after the Civil War, see James H. Cassedy, "The Registration Area and American Vital Statistics," *Bulletin of the History of Medicine*, 39 (1965), 221–231.

56. Levin S. Joynes, "Statistics of the Mortality of Baltimore, during a period of fourteen years, from 1836 to 1849 (inclusive)," *American Journal of the Medical Sciences*, n.s. 20 (1850), 313. See also "Partial Report upon a Uniform System of Registration of Births, Marriages, and Deaths, and the Causes of Death," *Transactions of the American Medical Association*, 9 (1856), 775; Franklin Tuthill, quoted in the editorial, "Registration of Births, Marriages, and Deaths," *Buffalo*

Medical Journal, 6 (1850 / 51), 691; Franklin Tuthill, "Registration," pp. 11–12; unsigned review, *Charleston Medical Journal*, 4 (1849), 475.

9. Sanitary Fact-Finding in the City

1. John H. Griscom, *Sanitary Legislation, Past and Future* (New York: Jones, 1861), p. 10.

2. Wilson Jewell, "Annual Report of the Committee on Public Hygiene," *Medical Examiner*, n.s. 8 (1852), 1–3.

3. Lewis Rogers, "A Lecture on Sanitary Reform," *Western Journal of Medicine and Surgery*, 3rd ser., 8 (1851), 512, 524.

4. Rogers, "Sanitary Reform," p. 507. For another expression of the crucial role of statistics, see "Public Health," *Buffalo Medical Journal*, 3 (1847 / 48), 757.

5. Prominent among American physicians who expressed conspicuous utilitarian views were Edward Jarvis and Jacob Bigelow, along with most of those who had public health responsibilities.

6. Rogers, "Sanitary Reform," pp. 512–513.

7. William Farr, "Letter to the Registrar-General," *Second Annual Report of the Registrar-General of Births, Deaths, and Marriages in England* (London: Her Majesty's Stationery Office, 1840), pp. 79–85. See also unsigned reviews, *New York Journal of Medicine*, 2 (1844), 218; ibid., 3 (1844), 66; "Diseases of Towns and of the Open Country," *Boston Medical and Surgical Journal*, 24 (1841), 26–29.

8. Chadwick's report was the largest part of a volume entitled *Report to her Majesty's Principal Secretary of State for the Home Department, from the Poor Law Commissioners, on an Inquiry into the Sanitary Condition of the Labouring Population of Great Britain* (London: Her Majesty's Stationery Office, 1842).

9. Charles Dickens, *American Notes, and Pictures from Italy* (London: Oxford University Press, 1957), pp. 251–252.

10. See, for instance, "Sanitary Condition of Philadelphia, *American Journal of the Medical Sciences*, n.s. 9 (1845), 522–523; "A Curious Statement," *Botanico-Medical Recorder*, 12 (1843 / 44), 390–391.

11. John Bell, *On Regimen and Longevity* (Philadelphia: Haswell & Johnson, 1842), pp. 22–23. See also [Samuel Forry], "On the Relative Proportion of Centenarians, of Deaf and Dumb, of Blind, and of Insane, in the Races of European and African Origin, as shown by the Censuses of the United States," *New York Journal of Medicine*, 2 (1844), 319; and "Registration of Marriages and Births," *Western Journal of Medicine and Surgery*, 3d ser., 2 (1848), 64–65.

12. "Registration of Births," *Medical Examiner*, n.s. 7 (1851), 266.

13. "Tenements of the Poor," *Boston Medical and Surgical Journal,* 35 (1847), 480–481.

14. G. E. [Gouverneur Emerson], review, *American Journal of the Medical Sciences,* n.s. 9 (1845), 396–401.

15. City of New York, Board of Aldermen, *Annual Report of the City Inspector, for the Year 1844,* Document no. 63. John H. Griscom, *The Sanitary Condition of the Laboring Population of New York* (New York: Harper, 1845).

16. In 1853, Griscom estimated that the 151,449 cases of disease treated at about twenty city institutions, mostly gratuitously, cost the city $745,458. Of 169 physicians working for these charities, 36 received board and room and 30 received salaries, while 103 received no financial compensation. John H. Griscom, *Anniversary Discourse before the New York Academy of Medicine* (New York: Graighead, 1855). See also *Western Journal of Medicine and Surgery,* 4th ser., 3 (1855), 283.

17. Griscom, *Sanitary Condition of the Laboring Population of New York,* passim.

18. Review, *New York Journal of Medicine,* 1 (1843), 88.

19. "Public Hygiene—Cellar Population," *New York Journal of Medicine,* n.s. 5 (1850), 120–122; "Ventilation: Subterranean Tenements," *American Phrenological Journal,* 17 (1853), 130.

20. *Transactions of the American Medical Association,* 1 (1848), 310; "First Report, Committee on Public Hygiene," ibid., 2 (1849), 431–654.

21. E. J. [Edward Jarvis], review, *American Journal of the Medical Sciences,* n.s. 9 (1845), 387; Alexander H. Stevens, "Annual Address," *Transactions of the Medical Society of the State of New York,* (1850), p. 19.

22. Lemuel Shattuck, "Letter to the Secretary," Dec. 12, 1845, in John G. Palfrey, *Fourth Annual Report to the Legislature, relating to the Registry and Returns of Birth, Marriages, and Deaths in Massachusetts, for the Year ending April 30th, 1845* (Boston: Dutton & Wentworth, 1845), p. 30. See also E. J. [Edward Jarvis], review, *American Journal of the Medical Sciences,* n.s. 9 (1845), 387.

23. E. J. [Edward Jarvis], "Report of the Sanitary Commission of Massachusetts," *Boston Medical and Surgical Journal,* 44 (1851), 89–90; Lemuel Shattuck, *Memorials of the Descendants of William Shattuck* (Boston: Dutton & Wentworth, 1855), p. 309; Jarvis, manuscript autobiography, Jarvis Papers, Harvard University. The full title of Shattuck's report is: *Report of a General Plan for the Promotion of Public and Personal Health, devised, prepared and recommended by the Commis-*

sioners appointed under a Resolve of the Legislature of Massachusetts, relating to a Sanitary Survey of the State (Boston: Dutton & Wentworth, 1850).

24. "Report of the Standing Committee on Medical Literature," *Transactions of the American Medical Association,* 4 (1851), 487.

25. For subsequent developments leading to formation of the Massachusetts Board of Health, see Barbara Gutmann Rosenkrantz, *Public Health and the State: Changing Views in Massachusetts, 1842–1936* (Cambridge, Mass.: Harvard University Press, 1972).

26. Jarvis to Chadwick, June 10, 1853, Jarvis Letterbook 3, Jarvis Papers, Harvard University.

27. "First Report of the Committee on Public Hygiene of the American Medical Association," *Transactions of the American Medical Association,* 2 (1849), 433–434.

28. John B. Blake, "Lemuel Shattuck and the Boston Water Supply," *Bulletin of the History of Medicine,* 29 (1955), 554–562.

29. "First Report, Committee on Public Hygiene," pp. 645–649.

30. Walter Channing, *An Address on the Prevention of Pauperism* (Boston: Christian World, 1843), pp. 44, 71, 83.

31. "Pure Water," editorial in *Boston Medical and Surgical Journal,* 7 (1833), 386–387.

32. Although 7 percent of the total deaths from 1807–1816 were attributed to this disease, only 5 percent were so caused between 1835 and 1844. "Importance of Pure Water in Cities," *Boston Medical and Surgical Journal,* 36 (1847), 409–412.

33. Griscom, *Sanitary Condition of New York,* pp. 44–55.

34. Shattuck, *Report of a General Plan,* passim.

35. Dr. Thomas W. Webb prepared the first state registration report for the committee. During the 1860s and 1870s Charles W. Parsons and Edwin M. Snow did much of this work. For a history of early registration in Rhode Island, see *First Report to the General Assembly of Rhode Island Relative to the Registry and Returns of Births, Marriages, and Deaths, and of Divorces in the State, for the year ending May 31, 1853* (Providence: Sayles, Miller, & Simons, 1854), pp. 1–60. See also review, *Boston Medical and Surgical Journal,* 56 (1857), 64–66.

36. E. M. Snow, *Statistics and Causes of Asiatic Cholera,* City Document no. 5 (Providence: 1855).

37. Charles W. Parsons, *Second Report to the General Assembly of Rhode Island relative to the Registry and Returns of Births, Marriages and Deaths of the State, 1853–1854* (Providence: Greene, 1856). Along

with his public health activities, Snow also conducted the census of Providence in 1855 and in several later years.

38. *City Registrar's Report on the Births, Marriages and Deaths, in the City of Providence, During the Year Ending December 31, 1855* (Providence: Knowles, Anthony, 1856). See also *City Registrar's Report . . . 1856; First Annual Report of the Superintendent of Health of the City of Providence for the Year Ending July 1, 1857, City Document no. 5* (Providence, 1857).

39. "Physicians' Certificates of the Causes of Death," *Boston Medical and Surgical Journal,* 61 (1859 / 60), 147–148; Wilson Jewell, "Report on Meteorology and Epidemics," *American Journal of the Medical Sciences,* n.s. 39 (1860), 389–390. For a further account of Snow's career, see James H. Cassedy, "Edwin Miller Snow: An Important American Public Health Pioneer," *Bulletin of the History of Medicine,* 35 (1961), 156–162.

40. Progressively better statistics became available on shipping, immigration, and diseases and mortality on shipboard, but studies specifically linking these data with the spread of given diseases on shore were not vigorously conducted.

41. *Minutes of the Proceedings of the Quarantine Convention held in Philadelphia . . . May 13–15, 1857* (Philadelphia: Crissy & Markley, 1857); *Minutes and Proceedings of the Second Annual Meeting of the Quarantine and Sanitary Convention, Baltimore, 1858* (Baltimore: Toy, 1858); *Proceedings and Debates, Third National Quarantine and Sanitary Convention, New York, 1859* (New York: Jones, 1859); *Proceedings and Debates, Fourth National Quarantine and Sanitary Convention, Boston, 1860* (Boston: Rand & Avery, 1860).

42. P. I. Wetmore, "Remarks," in *Proceedings and Debates, Third . . . Convention,* p. 107.

43. The New York State Medical Society, after some discussion, in January 1858 launched a plan to register medical and surgical cases, but few physicians cooperated. Joseph M. Smith, "Report on the Medical Topography and Epidemics of the State of New York," *Transactions of the American Medical Association,* 13 (1860), 172–174.

44. John Bell, *Report on the Importance and Economy of Sanitary Measures to Cities* (New York: Jones, 1859), pp. 198–199.

10. The Computing Mind at Midcentury

1. Oliver Wendell Holmes, "Currents and Counter-currents in Medical Science," in his *Medical Essays, 1842–1882* (Boston: Houghton Mifflin, 1889), p. 180.

2. Joseph LeConte, "On the Science of Medicine and the causes which have retarded its progress," *Southern Medical and Surgical Journal,* n.s. 6 (1850), 458.

3. Edward Jarvis, manuscript autobiography, and Jarvis to William Farr, Nov. 25, 1860, Jarvis Letterbook 6, Jarvis Papers, Harvard University.

4. Oliver Wendell Holmes, *Poet at the Breakfast Table,* in *Works of Oliver Wendell Holmes* (Boston: Houghton Mifflin, 1892), pp. 60, 168, 170.

5. Ralph Waldo Emerson, *Representative Selections,* ed. Frederic I. Carpenter (New York: American Book, 1934): "Experience," p. 183; "Politics," p. 198; "Fate," pp. 324–325; "Civilization," pp. 361–363.

6. Henry Jacob Bigelow, "Fragments of Science and Art," in his *Surgical Anaesthesia: Addresses and Other Papers* (Boston: Little, Brown, 1900), p. 180.

7. Oliver Wendell Holmes, "Currents and Counter-currents in Medical Science," in his *Medical Essays, 1842–1882* (Boston: Houghton Mifflin, 1889), pp. 193–194.

8. J. C. Simonds to Edward Jarvis, Nov. 5, 1852, Jarvis Papers, Harvard University; "Introductory Address," *New Orleans Medical Journal,* 1 (1844 / 45), i–vi.

9. Bennet Dowler, "Report of the Committee on Medical Sciences," *Transactions of the American Medical Association,* 4 (1851), 72–74. See also The Editor [Amariah Brigham], "Statistics of Insanity," *American Journal of Insanity,* 6 (1849 / 50), 141–145; editorial, "Statistics of Insanity," ibid., 18 (1861 / 62), 1–14; Bennet Dowler, "Postmortem Researches," *Western Journal of Medicine and Surgery,* 7 (1843), 241–245.

10. James L. Cabell, "Report of the Committee on Medical Education," *Transactions of the American Medical Association,* 7 (1854), 76.

11. G. C. S., review, *American Journal of the Medical Sciences,* n.s. 10 (1845), 143.

12. David W. Cheever, "The Value and the Fallacy of Statistics in the Observation of Disease," *Boston Medical and Surgical Journal* 63 (1860 / 61), 143, 449–456, 476–483, 497–503, 512–517, 535–541.

13. Review, *Buffalo Medical Journal,* 2 (1846 / 47), 11–22; C. A. L. [Charles A. Lee], review, *New York Journal of Medicine,* 4 (1845), 65–82; review, *Medical Examiner,* 7 (1844), 246–247; E. Leigh, "The Philosophy of Medical Science, considered with special reference to Dr. Elisha Bartlett's 'Essay on the Philosophy of Medical Science,' "

Boston Medical and Surgical Journal, 48 (1853), 69–74, 89–95, 115–121.

14. Jacob Bigelow, "Brief Exposition of Rational Medicine," in his *Modern Inquiries: Classical, Professional, and Miscellaneous* (Boston: Little Brown, 1867), pp. 244 ff.

15. Henry Jacob Bigelow, "Fragments," pp. 175–221.

16. A. S. [Stillé], review, *American Journal of the Medical Sciences,* n.s. 11 (1846), 405–408.

17. Austin Flint, "Lecture, Introductory to the Study of the Principles and Practice of Medicine," *Buffalo Medical Journal,* 5 (1849 / 50), 203.

18. L. A. Bertillon, *Conclusions statistiques contre les détracteurs de la vaccine; précédes d'un essai sur la méthode statistique appliquée à l'étude de l'homme* (Paris: Masson, 1857), p. vii; William Van Pelt, "Annual Address of the President of the Erie County Medical Society," *Buffalo Medical Journal,* 12 (1856 / 57), 554–557; Stanford Chaillé, "Anniversary Oration of the Physico-Medical Society," *New Orleans Medical and Surgical Journal,* 12 (1855 / 56), 579–591.

19. "Mathematics," *American Phrenological Journal,* 15 (1852), 127–128.

20. John William Draper, *Human Physiology, Statical and Dynamical* (New York: Harper, 1856). See also Bennet Dowler, "Experimental Researches upon Febrile Caloricity, both before and after death—Post-Mortem Fever," *Western Journal of Medicine and Surgery,* n.s. 1 (1844), 469–501; Bennet Dowler, "Researches into Animal Heat," *New Orleans Medical and Surgical Journal,* 17 (1860), 199–203; Bennet Dowler, "Critical and Speculative Researches on the Fundamental Principles of Subjective Science in connection with Medical and Experimental Investigations, with Remarks on the Present State of Medicine," ibid., 15 (1858), 39–64.

21. Joseph Jones, "Observations on Malarial Fever," *Southern Medical and Surgical Journal,* n.s. 14 (1858), 375–376.

22. Austin Flint, "Observations on the Pathological Relations of the Medulla Spinalis," *American Journal of the Medical Sciences,* n.s. 7 (1844), 269–271; Austin Flint, letters from Paris, *Buffalo Medical Journal,* 10 (1854), pp. 107–114, 129–142, 193–197, 321–332, 412–419; Claude Bernard, *An Introduction to the Study of Experimental Medicine* (New York: Dover, 1957), pp. 134–135; Hebbel E. Hoff, Roger Guillemin, and Edvart Sakiz, "Claude Bernard on Experimental Medicine—Some Unpublished Notes," *Perspectives in Biology and Medicine,* 8 (1964), 34–35.

Index